Internal Rotation and Inversion

An Introduction to Large Amplitude Motions in Molecules

Internal Rotation and Inversion

An Introduction to Large Amplitude
Motions in Molecules

**DAVID G. LISTER, JOHN N. MACDONALD
and NOEL L. OWEN**

School of Physical and Molecular Sciences, University College
of North Wales, Bangor, Wales

1978

ACADEMIC PRESS
London   New York   San Francisco

A Subsidiary of Harcourt Brace Jovanovich, Publishers

ACADEMIC PRESS INC. (LONDON) LTD.
24/28 Oval Road, London NW1 70X

United States Edition published by
ACADEMIC PRESS INC.
111 Fifth Avenue, New York, New York 1003

Library of Congress Catalog Card No: 77–15322
ISBN: 0–12–452250–5

Printed in Great Britain by
Page Bros (Norwich) Ltd,
Mile Cross Lane, Norwich

Preface

The motives of authors of scientific monographs and text books are as many and varied as the authors themselves. In our case we chose to write this book largely because we are interested in the subject. Also it seemed to us that although many admirable and detailed accounts of aspects of this subject are available in specialist texts and publications, there does not seem to be a book available at the present time which gathers together for the non-specialist the now very extensive, and in some cases very detailed, information concerning large amplitude motions in molecules. We hope that this book may prove useful to advanced undergraduates who find molecular structure interesting and to graduate students and other research workers who may be working in fields where the detailed features of molecules are important but perhaps not all consuming.

We have not attempted to produce a comprehensive text book so we have deliberately chosen to include suggestions for further reading at the end of each chapter so that those who are interested can proceed to a more detailed account of any particular topic. In addition we have included a selected number of key references for each chapter. We have also omitted certain topics which, although related to large amplitude motions in molecules, really lie beyond the scope which we have set ourselves.

We have attempted to restrict to a minimum the mathematical demands on the reader and several sections of the book, notably Chapters 1, 3 and 9 require no prerequisite mathematical knowledge. However, we necessarily assume some familiarity with basic chemical and physical notions of molecular structure and also a modest acquaintance with the ideas of quantum mechanics but only to the extent to which they are covered in most undergraduate chemistry courses. In the chapter dealing with Molecular Energy Levels we have included an outline of some of the more important applications of the principles of quantum mechanics to the study of molecular dynamics.

The study of large amplitude, hindered movements of atoms in molecules may seem to be a particularly specialized and narrow field of activity but consequences of these motions are felt in widely differing aspects of molecular science. For example the ammonia maser, the forerunner of the now common laser, derives its remarkable property from the well known "umbrella" inversion of the molecule, and the detection, using radio astronomical techniques, of a spectral line associated with this inversion was one of the earliest pieces of evidence for the existence of polyatomic molecules in interstellar space. Also, the process of forming helices or coils for biologically important molecules is governed to some considerable extent by the ability of groups of atoms to arrange themselves by rotation around covalent bonds. The subject matter of the book cuts across most of the traditional boundaries of chemistry and physics and it is significant that much of the original work in this field is found in journals devoted to interdisciplinary areas such as molecular structure, molecular spectroscopy and chemical physics. It is a field to which both chemists and physicists alike have contributed significantly.

We take full responsibility for any errors which may be present in the text, but we would like to thank Professor J Sheridan for reading the manuscript and Dr J R Turvey for reading parts of it and for suggestions which have improved the text. All three of us have, at one time or another, worked closely with Professor Sheridan in the field of molecular structure and we acknowledge with gratitude his help and support. To Margaret Macdonald go our sincere thanks for the excellent way that she produced good typescript out of poor manuscript, and for her patient endurance over a lengthy period of time.

We are also very grateful to the following for permission to reproduce material from their published work; Dr H W Harrington, Professor E. Hirota, Dr D. A. Rees and Dr O. L. Stiefvater.

January, 1978

D. G. Lister
J. N. Macdonald
N. L. Owen

Foreword

The increase, inside a generation, in our knowledge of large amplitude motions within molecules is as striking as any of the other transformations in physico-chemical knowledge in the same period. The impression is easily gained, however, that among chemists, whose subject now depends overwhelmingly on subtleties of molecular shapes, and indeed among some physicists, the extent of the increase in our understanding of the energetics of changes in molecular conformations is not as well appreciated as it should be. Summarizing the findings in this area is a difficult task in which a compromise must be made between the seemingly boundless mathematical expression of rigour, and the fascinating but inadequate mechanistic descriptions of the phenomena which link the subject to more familiar molecular behaviour. My three colleagues who have written this book, all of whom have made their own distinctive contributions to this field, have convincingly steered such a course, in which, while we are not allowed to ignore the value of a proper fundamental approach, we can appreciate the real contributions of such studies to what ought to be known by every scientist who deals with molecules. It is an account which should serve well the undergraduate or graduate student in chemistry or physics, who wishes to understand the advances in this subject. I commend it as a move towards a more general appreciation of the value of the advances in this field.

J. Sheridan

Contents

ix

1
Molecules, Isomerism and Large Amplitude Vibrations

1.1 Introduction

Atoms in molecules are held together by electrostatic interactions between electrons and nuclei. Chemical bonds are a manifestation of these interactions, and both the structure of molecules and the nature of the vibrational motions of the constituent atoms are governed by the types of bonds present in a molecule. The relationship between structure and vibration is seen clearly in diatomic molecules where there is only one mode of vibration, the bond stretching mode, and for such a molecule the chemical binding energy varies with internuclear distance in the manner shown in Fig. 1.1. This curve also represents the potential energy function for the vibration of the nuclei and it governs the separation of the vibrational energy levels and the classical amplitudes of vibration.

The minimum of the curve corresponds to the equilibrium internuclear distance (r_e) and the strength of the chemical bond is reflected in the magnitude of the dissociation energy (D_e) which is the energy required to split the molecule into its constituent atoms. A measure of how difficult it is to distort the molecule from its equilibrium configuration is expressed by the bond force constant (k) which is related to the potential energy by the relationship,

$$k = \left(\frac{\partial^2 V}{\partial r^2}\right)_{r=r_e} \tag{1.1}$$

V represents the potential energy and r is the separation between the nuclei. Table 1.1 summarizes the molecular properties of three simple diatomic molecules and the correlation with the type of chemical bonding found in these molecules is evident.

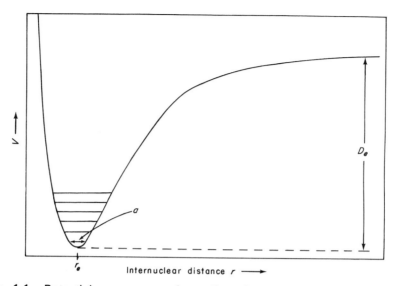

Fig. 1.1 Potential energy curve for a diatomic molecule showing the equilibrium internuclear distance, equilibrium dissociation energy and the classical amplitudes of vibration (a).

Analyses of polyatomic molecules are more complex since the chemical binding energy depends on several internuclear distances and upon the relative geometry of the constituent atoms. However, the bonding can be represented by a multidimensional potential energy surface where minima in the surface correspond to stable arrangements of atoms. The relationships between the shapes of molecules and their physical and chemical properties have fascinated many scientists, and one of the most exciting and challenging fields of study in chemistry and physics over the past fifty years or so has been

Table 1.1   Some molecular parameters for three diatomic molecules

	N≡N	O=O	F—F
r_e/pm	109·4	120·7	143·5
D_e/J	$1·59 \times 10^{-18}$	$0·84 \times 10^{-18}$	0.22×10^{-18}
k/Nm^{-1}	2240	1140	445
u/pm[a]	3·2	3·7	4·7

[a] "u" is the "root-mean-square" amplitude of the molecular vibration in the ground ($v = 0$) state.

the development of new techniques to study in depth the spatial arrangements of atoms within molecules.

1.2 Isomerism

Certain molecular configurations are much more stable than others; consequently, when a molecule is deformed and the atoms displaced from their equilibrium positions, the total potential energy of the molecule increases. A polyatomic molecule, however, may have several different sets of minima in its potential energy surface, each set corresponding to a stable arrangement of atoms. If the potential energy barriers to the rearrangement of atoms are very high then the different molecular species may exist as separate entities and isomerism occurs. Some examples are shown in Fig. 1.2, viz: methyl cyanide and methyl isocyanide (Fig. 1.2(a)) and *cis* and *trans* 1,2-dichloroethylenes (Fig. 1.2(b)).

Other molecules have barriers which are sufficiently high to give the different molecular species a separate existence but not high enough to prevent interconversion of the species by passage over the barriers in a classical manner or through them by quantum mechanical tunnelling. When interconversion of such isomers takes place by rotation of groups of atoms about a chemical bond then they are called *rotational* or *conformational isomers*. The first compounds for which rotational isomerism was shown to occur were substituted derivatives of diphenyl, where steric interactions

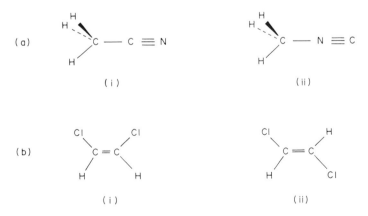

Fig. 1.2   Isomeric forms of molecules, (a) (i) methyl cyanide and (ii) methyl isocyanide, and (b) (i) *cis* and (ii) *trans* 1,2-dichloroethylene.

Fig. 1.3   Resolution of the two optically active isomers of 2,2'-dinitrodiphenyl-6,6'-dicarboxylic acid represented one of the first pieces of experimental evidence for the occurrence of hindered internal rotation in molecules.

between the side groups are so large that interconversion of conformers at normal temperatures is not possible (Fig. 1.3).[1] Here the two stable conformations may be resolved as separate *optical isomers*. In general, however, individual rotational isomers cannot be isolated using standard chemical or physical methods except under very special circumstances, e.g. at very low temperatures. Under these latter conditions the vibrational energy of the molecules is so low that the potential energy barriers to interconversion become virtually insurmountable. Rotational isomers (or rotamers as they are often called) may be physically indistinguishable or they may represent quite different geometric structures. An example of one indistinguishable pair of isomers is found in phenol (Fig. 1.4(a)), where rotation of the hydrogen atom

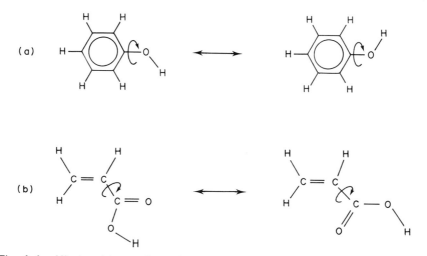

Fig. 1.4   Hindered internal rotation in (a) phenol, giving rise to two identical rotational isomers, and (b) acrylic acid, giving rise to *trans* and *cis* rotamers.

about the C—O bond produces identical conformers every 180°. On the other hand, internal rotation about the central C—C bond in acrylic acid produces different molecular conformations, the two most stable species being shown in Fig. 1.4(b). Interconversion of stable molecular conformations may also be brought about by inversion of a molecule as exemplified by ammonia (Fig. 1.5).

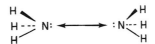

Fig. 1.5   The inversion mode in ammonia.

This book is largely concerned with aspects of internal rotation, inversion and conformational isomerism, the energy differences between rotamers and the potential barriers separating them. A variety of different experimental and theoretical techniques are used to obtain and analyse structural data for molecules, and it is our hope that this book will serve as an introduction to some of the facets of this field of study.

1.3 Potential Energies Associated with Internal Rotation

The origin and nature of the hindering potential barrier to internal rotation is a subject of great interest for both experimentalists and theorists. (A brief review of the main theories propounded in connection with internal rotation is included in Chapter 4.) The magnitude of these potential barriers can vary from a few $J\,mol^{-1}$ up to several hundreds of $kJ\,mol^{-1}$, the lower limit representing almost free rotation about a single bond, and the upper limit corresponding to the energy required to twist a normally rigid double bond.

If the potential energy of a molecule capable of exhibiting rotational isomerism is expressed as a function of the angle of internal rotation (α), the resulting graph of potential energy against α shows a series of maxima and minima. For example, in acetaldehyde, the potential energy associated with internal rotation about the central C—C bond varies with angle of rotation in the manner illustrated in Fig. 1.6. Also shown in the figure is the equilibrium conformation of acetaldehyde. During one complete revolution of the CH_3 group (i.e. 360°), *three* equivalent stable conformations are generated (at 120°, 240° and 360°) and the associated potential function is said to be *threefold* symmetric. (The experimental methods which establish which

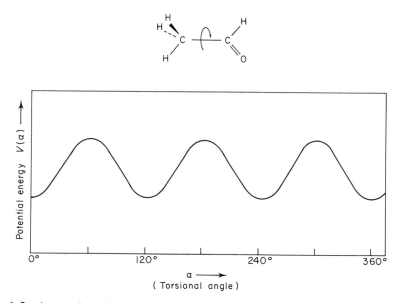

Fig. 1.6   Internal rotation in acetaldehyde showing the most stable conformation of the molecule and the shape of the potential function governing internal rotation.

conformations are stable are outlined later in this chapter and some of them are discussed in greater detail in Chapter 3.) In the equilibrium molecular conformation of acetaldehyde the C—H bonds on the carbon atoms are staggered. The three conformations obtained during one complete revolution of the methyl group are exactly equivalent to each other and there is no way of distinguishing one from another. Provided that the potential barrier between the minima is not large compared with the average kinetic energy of a molecule, then the three hydrogen atoms will exchange their positions by the process of internal rotation or by quantum mechanical tunnelling through the barrier. For acetaldehyde the difference in energy between the minima and the maxima of the potential energy curve has been found experimentally to be 4.89 kJ mol^{-1} ($1.17 \text{ kcal mol}^{-1}$),† and since the average kinetic energy at ambient temperature is about 2.5 kJ mol^{-1}, interconversion of the hydrogen atoms of the methyl group occurs fairly readily. It is interesting to note at this juncture that the barriers to internal rotation for related molecules such as acetyl fluoride and acetyl cyanide are very similar to the value found for

---

† In the S.I. system of units (Système Internationale d'Unités) energy is expressed as J mol^{-1}. It should, however, be noted that the vast majority of potential barriers and energy differences quoted in the literature are expressed in units of cal mol^{-1}.
($1 \text{ cal mol}^{-1} \equiv 4.184 \text{ J mol}^{-1} \equiv 0.3498 \text{ cm}^{-1}$)

acetaldehyde (4.36 and 5.32 kJ mol^{-1} respectively). Where the symmetric methyl group remains unchanged, but where different substituents are placed on the asymmetric framework about which it rotates, as in the above examples, the overall shape of the potential energy curve remains essentially the same. However, quite different potential energy curves are obtained if the asymmetric framework changes its form drastically—as for example in toluene. In toluene (Fig. 1.7(a)), the framework about which the methyl group rotates has more symmetry than in acetaldehyde. Figure 1.7(b) illustrates that as any two hydrogen atoms of the methyl group stagger a C—C bond on the framework so the other hydrogen eclipses an identical C—C bond on the opposite side.

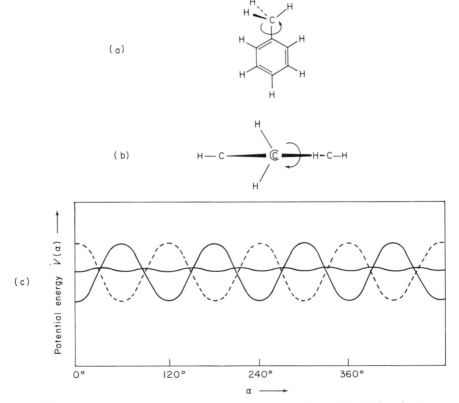

Fig. 1.7 Internal rotation in (a) toluene, showing (b) a stable molecular conformation viewed along the CH$_3$—C$_6$H$_5$ bond, and (c) the cancellation of two threefold potential functions 180° out of phase arising from rotation of a top with threefold symmetry against a twofold symmetric frame. A small sixfold potential function to internal rotation is found for toluene.

One way of viewing such a potential energy curve is to consider it to be composed of two identical threefold potentials exactly out-of-phase with each other (Fig. 1.7(c)). The result is a cancellation of the threefold potential energy contributions. The barrier for toluene must therefore be composed of higher terms than threefold in the Fourier expansion of the potential energy (see Section 2.6.1), and there are *six* identical minima in the potential energy curve for every complete revolution of the methyl group. The experimentally measured barrier for toluene is 58 J mol^{-1}, indicating the fact that potential functions with higher symmetry generally have lower barriers to internal rotation.

If we retain the aldehyde framework as in acetaldehyde, but alter the nature of the adjacent group from CH_3 to CH_2CH_3, then the resulting potential energy curve for internal rotation about the C_2H_5—CHO bond in propionaldehyde is shown in Fig. 1.8.

The most stable conformer has the four heavy atoms coplanar, with the oxygen atom *cis* to the methyl group, while other stable conformations correspond to *gauche* structures (approximately 120° and 240° on the diagram shown in Fig. 1.8). The *cis* and *gauche* conformations are examples of

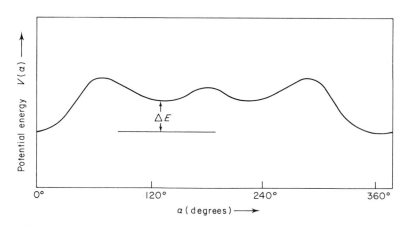

Fig. 1.8   Internal rotation in propionaldehyde showing the most stable conformation of the molecule and the approximate shape of the potential function to internal rotation about C_2H_5—CHO bond.

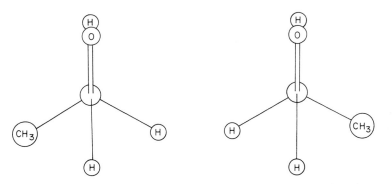

Fig. 1.9. Two *gauche* rotamers of propionaldehyde; these are mirror images of each other.

rotational isomers, and, in addition, the two equivalent *gauche* conformations are mirror images of one another and thus represent *optical isomers* (Fig. 1.9). (At normal temperatures, no optical activity would be expected owing to the ease of interconversion of the rotamers through rotation around the C—C bond.) Although they have relatively short lifetimes when compared with geometric isomers, rotational isomers can be identified and distinguished by several experimental techniques.

In addition to the height of the potential barrier, the energy difference between rotational isomers (i.e. the difference in energy between the minima of a potential energy curve, as for example in *cis* and *gauche* propionaldehyde (Fig. 1.8)) also represents an important, experimentally determinable parameter in internal rotation studies. This energy difference governs the magnitude of the equilibrium population ratio of the two rotational isomers at any temperature, via the Boltzmann expression,

$$\frac{n_1}{n_2} = g\, e^{-\Delta E/kT} \tag{1.2}$$

where ΔE is the energy difference, k is the Boltzmann constant and g is the statistical weight ratio of the rotamers (e.g. in propionaldehyde, $g = 2$ in favour of the *gauche* conformer). Figure 1.10 illustrates for propionaldehyde how the percentage of the more stable rotamer in equilibrium with the second rotamer would vary with the energy difference between them at certain temperatures.

The ease of interconversion of rotational isomers is governed by the height of the potential barrier separating them, but the equilibrium ratio of rotamers

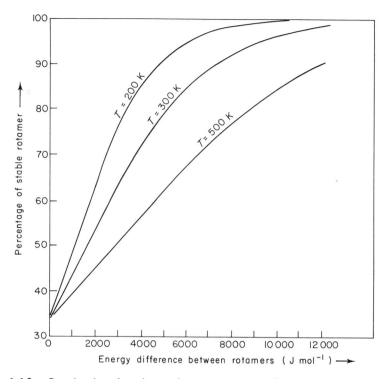

Fig. 1.10 Graph showing how the percentage of the thermodynamically more stable rotamer of a molecule such as propionaldehyde, with respect to the less stable, doubly degenerate form $(g = 2)$ varies with the energy difference between the rotamers at certain temperatures.

is a function only of the energy difference between them. If hindered internal rotation is treated clasically as a simple rate process the rate constant k for internal rotation is given by the Arrhenius equation,

$$k = A\, e^{-E_A/kT} \tag{1.3}$$

where the activation energy E_A corresponds to the height of the potential barrier to internal rotation. Several attempts have been made to correlate the pre-exponential factor A with the frequency of torsional oscillation since they may be shown to be of the same order of magnitude, but it is unlikely that a purely classical description can successfully account for all the features of a phenomenon which may be partially interpreted in terms of quantum mechanical tunnelling through the potential barrier.

1.4 The Experimental Approach

Many different experimental methods have been used to study large amplitude vibrations in molecules and descriptions of the four most important methods are given in Chapter 3. It will be of value, however, in this introductory chapter, to outline the basic principles of some of the methods in a general manner. Table 1.2 summarizes the methods and the main parameters which can be derived using them.

In studies of internal rotation and other large amplitude motions in molecules, the key parameters of interest include the following,

(a) The number of different stable rotational isomers in equilibrium and the energy differences between them.
(b) The preferred conformation (and the molecular geometry, if possible) of stable rotameric forms.
(c) The shapes of the potential energy curves for the large amplitude motions, and the heights of barriers to inversion, ring puckering and internal rotation.

The main physical methods used for studying molecular motions and obtaining values for these parameters may be classified into the following broad categories: (i) spectroscopic, (ii) diffraction, (iii) relaxation and (iv) classical.

1.4.1 Spectroscopic methods

All spectroscopic methods involve the interaction of electromagnetic radiation with matter and the subsequent absorption (or emission) of discrete quanta of energy. The most useful methods in this category include infrared, Raman and microwave spectroscopy, and to a certain extent nuclear magnetic resonance spectroscopy, although the latter has, in the main, been most usefully employed as a relaxation technique.

At ambient, and below ambient, temperatures, most molecules reside in their lowest vibrational states, and so infrared and Raman data tend to refer to molecules in their ground vibrational states ($v = 0$). The normal vibrations of a molecule depend on the molecular geometry and so, in principle, different rotamers can give rise to quite different absorption lines. In practice, however, many absorptions are coincident and only a few lines are doubled by rotational isomerism. The use of both Raman and infrared data in concert often enables the gross molecular conformations of rotamers to be deduced,

Table 1.2   Methods of studying internal motions in molecules and the important derived molecular parameters

Method	Principle	Parameters determined
1. Infrared and Raman spectroscopy	(a) frequency measurements of torsional modes	potential barriers
	(b) relative intensities of absorptions with changes in temperature	energy difference between rotamers
	(c) shapes of absorption bands	(i) molecular conformation (gases)
		(ii) potential barriers (liquids and solids)
	(d) frequency measurement of rotational fine structure	molecular structure (gas phase, small molecules only)
2. Microwave spectroscopy	(a) frequency measurements	(i) molecular conformation and precise molecular structure
		(ii) potential barriers
	(b) relative intensities of bands with or without changes in temperature	(i) potential barriers
		(ii) energy difference between vibrational states and/or between rotamers
3. Nuclear magnetic resonance spectroscopy	(a) line shape analysis	potential barriers
	(b) relaxation measurements	potential barriers
	(c) coupling constant analysis for liquid crystals	potential barriers

Method		Information
4. Electronic spectroscopy	(d) long range coupling constant measurements	molecular conformation
	analysis of vibrational structure of electronic transitions	(i) potential barriers (ii) rotational isomerism
5. Electron diffraction	(a) radial distribution patterns	molecular conformation
	(b) relative intensity measurements	energy difference between rotamers
	(c) synthesis of scattering curve	potential barriers
6. Neutron inelastic scattering	energy analysis of scattered neutrons	torsional frequencies and potential barriers
7. Ultrasonic relaxation	frequency measurements of sound absorption with varying temperature	(i) potential barriers (ii) energy difference between rotamers
8. Dipole moments	measurement of dipole moment as a function of temperature	energy difference between rotamers
9. Thermodynamic methods	comparison of experimental and calculated values of entropy and heat capacity	potential barriers
10. Electron spin resonance spectroscopy	line shape analysis	potential barriers of species with unpaired electrons
11. Kerr effect	measurement of optical anisotropy in an electric field	molecular conformation

and low frequency torsions and the frequencies of other large amplitude vibrations may be obtained directly for samples in solid, liquid or gaseous phases.

Microwave spectroscopy represents an immensely powerful technique for determining molecular structures but one which can be utilized only for polar vapours. The complex patterns of spectral lines arise from many different rotational transitions originating in the ground and other vibrational states. The lines from any one vibrational state are characterized by a unique set of rotational constants which themselves depend directly on the molecular structure (see Chapter 2). The effect of hindered internal rotation can often be observed on microwave spectra where the usual rigid rotor spectral lines may be broadened or split into two or more components. Largely as a consequence of the great accuracy with which absorption lines can be measured, microwave spectroscopy can produce quite precise values for the various parameters associated with large amplitude molecular motions.

For a substance capable of existing as different rotational isomers, separated by finite potential barriers, the various conformers may be identified spectroscopically as separate molecular species. In a classical sense, an instantaneous "snapshot" of the molecule is obtained provided that the mean lifetime of a given rotamer is long compared with the time taken for the interaction to occur with the radiation. Because the experimental data mostly refers to the stable equilibrium geometry of molecules (i.e. near the minimum of the potential energy wells) predictions relating to the heights of potential barriers are largely dependent on assumptions made about the shape of the potential energy curves.

1.4.2 Diffraction methods

Electrons, neutrons and X-rays are all diffracted by matter under different circumstances, and useful structural information related to molecules can be obtained from the respective diffraction patterns. X-rays are scattered primarily by the electrons in the molecule, and since X-rays have a high penetrating power the best diffraction patterns are obtained with solid crystals where vast numbers of scattering centres (i.e. atoms) are repeated at regular intervals. X-ray diffraction is not particularly well suited to the study of rotational isomerism since in the solid state, molecules are mostly confined to one particular stable conformation.

Neutron diffraction, which is more limited as a structural probe than X-ray diffraction, is best applied to locate hydrogen atoms in molecules. Its lack of charge enables a neutron to penetrate the outer electrons of an atom and the scattering centre in this instance is the atomic nucleus, and the scattering power does not increase systematically with atomic number. Hydrogen atoms are not well detected by X-rays and so neutron diffraction studies have been found to be a valuable complement to X-ray diffraction for the study of systems involving hydrogen bonds.

Neutrons have been used in the study of molecules over the past ten years in another quite different way.[2] The inelastic scattering properties of neutrons may be used to great advantage in estimating low frequency vibrations of compounds. An approximately monochromatic stream of low energy neutrons ($\sim 0.3 \text{ kJ mol}^{-1}$) is generated from a nuclear reactor, and the incoherent neutron scattering from the sample measured with a time-of-flight spectrometer. Both liquid and gaseous samples can be studied, and the scattering process may be compared to the more commonly used Raman photon scattering at optical frequencies, in that the scattered neutron gains a quantum of vibrational energy from the scattering molecule. One important difference between the two processes is that neutrons are not subject to optical selection rules, and so optically "forbidden" transitions, such as the torsion in ethane, may be observed directly. The technique has proved very useful for measuring large amplitude low frequency vibrations involving hydrogen atoms.

Diffraction of monochromatic electron beams by gaseous samples represents one of the most powerful methods of determining accurate molecular parameters, and the technique is described in greater detail in Chapter 3. Where rotational isomers are thermodynamically stable, they may be identified by electron diffraction as distinct molecular species, provided that some of the atom–atom distances within the various rotamers are significantly different.

The pattern obtained from diffraction experiments represents an average over all the molecules in all the different energy states at the operating temperature and over all possible molecular orientations, hence the derived molecular structures do not always correlate readily with those derived from spectroscopic studies.

1.4.3 Relaxation methods

The equilibrium between two rotational isomers may be regarded as the result of two competing unimolecular rate processes,

$$I \underset{k_{21}}{\overset{k_{12}}{\rightleftharpoons}} II$$

with a free energy diagram as shown in Fig. 1.11.

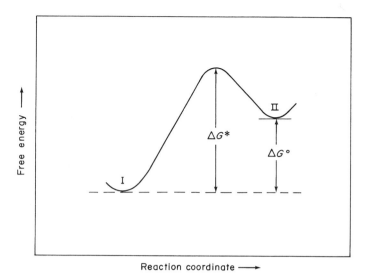

Fig. 1.11   Free energy diagram for the interconversion of two rotational isomers showing the respective equilibrium positions. ΔG^* is the free energy difference between the transition state and the more stable isomer.

A molecule in the conformation represented by I can convert to that represented by II by acquiring sufficient energy through collisions with other molecules to pass over the top of the barrier. This description is appropriate to liquids or solutions of molecules containing heavy groups and with high barriers to internal rotation or inversion. Quantum mechanical tunnelling through the barrier is not important in these circumstances.

If the equilibrium represented in Fig. 1.11 is disturbed by a sudden increase in temperature or pressure then the system will relax to its new equilibrium at a rate (R) given by,

$$R = k_{12}[I] - k_{21}[II]$$

where $[I]$ and $[II]$ are the concentrations of rotamers I and II. The *relaxation time* (τ) for the process is defined as the time taken for the system to return a fraction $1/e$ of the way to equilibrium, and may be expressed as,

$$\tau = \frac{1}{k_{12} + k_{21}} \qquad (1.4)$$

The appearance of the exponential constant (e) in the definition of τ results from relating the isomerization process to a first-order reaction scheme, whereby the concentration of the reactant $[C]$ at any time is related to the initial concentration $[C_0]$ by,

$$[C] = [C_0] e^{-t/\tau} \tag{1.5}$$

When $t = \tau$, $[C] = [C_0]/e$. The rate constant k_{12} may be related, through the Arrhenius equation, to the activation energy for the process,

$$k_{12} = A e^{-E_A/kT} \tag{1.6}$$

The activation energy is directly comparable to the potential energy barrier for the internal rotation. Since the equilibrium constant (K) may be expressed as the ratio of the forward and reverse rate constants,

$$K = k_{12}/k_{21}$$

equation (1.4) may be rewritten as,

$$k_{12} = \frac{1}{\tau} \left(\frac{K}{1 + K} \right) \tag{1.7}$$

Hence measurement of relaxation times provides a means of estimating barriers hindering internal rotation and other large amplitude vibrations.

The relaxation time is a measure of the mean lifetime of a particular molecule as either rotamer I or II, and molecules may be regarded as interconverting between the two rotamers at a frequency ω, where $\omega = 1/\tau$. For barrier heights in the range 10 to 50 kJ mol^{-1}, ω varies between 5 and 150 MHz. Nuclear magnetic resonance spectrometers generally operate at frequencies of 60 or 100 MHz and, as described in Chapter 3, the interconversion of rotamers with barriers greater than about 25 kJ mol^{-1} can have a drastic effect on nmr line shapes, and this can be used to measure relaxation times.

Ultrasonic techniques can also be used to measure the relaxation times associated with the interconversion of rotamers.[3] The propagation of an acoustic wave through a medium may be regarded as causing alternating compressions and rarefactions of the volume elements of the medium. (Acoustic waves are longitudinal in character as opposed to electromagnetic waves which are transverse.) Since the period of these fluctuations is small compared to the times required for thermal equilibration with the surroundings, the process is essentially adiabatic. Attenuation of the wave is caused by an exchange of energy with the propagating medium. When the frequency of the ultrasonic waves is comparable to $1/\tau$, an anomalous loss of energy and

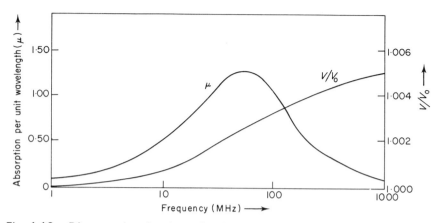

Fig. 1.12   Diagram showing how the absorption per unit wavelength (μ) and the relative change in velocity (v/v_0) change with frequency for a typical relaxation process.

change in velocity of sound in the medium occurs. The absorption per unit wavelength (μ) and the relative change in velocity (v/v_0) for a typical relaxation process are shown in Fig. 1.12. The maximum value of μ occurs at a frequency $f_c = 1/(2\pi\tau)$. Measurement of the variation of μ with temperature makes it possible to obtain thermodynamic information, while the temperature variation of f_c yields kinetic data. One disadvantage of the ultrasonic method is that the relaxation process affects the absorption and velocity over a wide frequency range (Fig. 1.12) and therefore it is difficult to resolve two or more processes unless their relaxation times differ considerably. The method is best suited for liquids, and may be used to determine potential barriers in the range 10–60 kJ mol^{-1}.

Another related relaxation technique, but one which has not been used very often to give quantitative information about internal rotation and other molecular motions, is dielectric relaxation. If a static or slowly varying electric field is applied to a liquid or gaseous sample of a polar substance the molecular dipoles become partially orientated in the direction of the electric field. Under these conditions molecules capable of internal rotation appear to have a dipole moment which is the time average over all of the possible molecular conformations. When the frequency of the applied electric field is comparable to $1/\tau$ the intramolecular dipoles do not attain equilibrium with the applied electric field. An absorption of electrical energy occurs and this is characterized by a decrease in the dielectric constant. The method has been found useful for detecting molecular motions in large molecules, including polymers.

Relaxation techniques differ from the other methods described in this chapter in that they actually measure the rate at which rotamers interconvert. However, they are mainly used to determine barriers in the liquid or solution phase where intermolecular interactions may be important. It may be argued that such measurements have more relevance to everyday chemical and life processes, although they are much more difficult to interpret than gas phase measurements.

1.4.4 Classical methods

Historically very important, the thermodynamic and dipole moment methods basically consist of accurately measuring thermodynamic parameters (such as enthalpies, entropies, etc.) or determining dipole moments by measuring capacities, refractive indices, etc., and then comparing these values to those calculated assuming certain molecular models. The results consequently reflect the validity of the model. These, and related methods (e.g. Kerr Constant studies), measure bulk properties of matter and the derived data refers to a statistically averaged molecular structure.

A detailed account of the application of both thermodynamic and dipole moment techniques to the study of internal rotation may be found in the monograph written by Mizushima (see ref. 1, Further Reading). Where the dipole moments of individual conformers are significantly different (e.g. 1,2-dichloroethane, $\mu(trans) = 0$; $\mu(gauche) = 8.7 \times 10^{-30}$ Cm)† measurement of the mean dipole moment of the equilibrium mixture as a function of temperature enables the energy difference between the rotamers to be evaluated.

Measurement of the optical anisotropy of a liquid or vapour sample leads to a value of the Kerr constant which is related to the molecular dipole moment and polarizability.[4] When a transparent isotropic substance (liquid or gas) is placed in a stationary electric field, reorientation of the molecular dipoles occurs, and for non-spherically symmetrical molecules the sample becomes optically anisotropic and doubly refracting. The Kerr constant is a function of the refractive indices of the emergent light in two directions at right angles and also of the applied electric field. Deductions concerning molecular conformations may be made by comparing the experimentally derived Kerr constants with the values calculated for various molecular conformations.

---

† The majority of dipole moment values quoted in the literature are expressed in debyes (D); 1 debye $= 10^{-18}$ esu cm $= 3.333 \times 10^{-30}$ Cm.

Further Reading

1. S. Mizushima. "Structure of Molecules and Internal Rotation". Academic Press, London and New York, 1954.
   The first of its kind, and although a little dated in some aspects, this still represents a very good introduction to hindered internal rotation.
2. E. L. Eliel, N. L. Allinger, S. J. Angyal and G. A. Morrison. "Conformational Analysis". Interscience, Chichester and New York 1965.
   A classic book on molecular conformation, concentrating largely on ring compounds.
3. W. J. Orville-Thomas (Ed.). "Internal Rotation in Molecules". Wiley, Chichester and New York, 1974.
   An edited volume incorporating details of many of the modern methods of studying internal rotation.

References

1. G. H. Christie and J. Kenner. *J. Chem. Soc.* **LXXI**, 614 (1922).
2. H. Boutin and S. Yip. "Molecular Spectroscopy with Neutrons". MIT Press, Cambridge, Mass, 1968.
3. S. M. Walker. Molecular acoustics and conformational behaviour, Chap. 9 *in* "Internal Rotation in Molecules" (W. J. Orville-Thomas, Ed.). Wiley, Chichester and New York, 1974.
4. R. J. W. Le Fèrve. Molecular refractivity and polarizability, Chap. 1 *in* "Advances in Physical Organic Chemistry" (V. Gold, Ed.), Vol. 3. Academic Press, London and New York, 1965.

2
Molecular Energy Levels

2.1 Introduction

Much of the material in later chapters requires some background knowledge of a variety of theoretical or mathematical techniques. In an introductory book it is impossible to do more than introduce the necessary language and point out how it is applied to the subject at hand. To deal with a topic such as internal rotation, which involves a range of different experimental techniques and methods of deriving molecular information, a certain flexibility of outlook is called for. In some circumstances, one must be prepared to adopt a very simple approach in order to be able to derive any information or, at the other extreme, it may be necessary to go into a very detailed treatment in order to discuss some of the finer points which may arise in the interpretation of experimental results. At some stage most problems require solution of the Schrödinger equation and this is usually done on a computer using matrix methods. This chapter therefore starts with some remarks about solving the Schrödinger equation. This is followed by a more detailed discussion of the molecular Hamiltonian operator and molecular vibrations. The mathematical formulation of the potential energy curves for internal rotation and other large amplitude motions will be considered and the solution of the Schrödinger equation will be illustrated by two important examples. Finally the rotational motion of molecules and the interaction between the vibrational and rotational motions will be briefly treated.

2.2 Solution of the Schrödinger Equation

The Schrödinger equation may be written in terms of the Hamiltonian operator (H) as,

$$H\psi = E\psi \qquad (2.1)$$

where the Hamiltonian operator is the sum of two operators (T and V) corresponding to the classical kinetic and potential energies. For a number of idealized systems, e.g. the harmonic oscillator, this equation may be solved exactly and these solutions may be used as a basis for obtaining the energy levels and wave functions for more complicated problems.[1] Often it is possible to write the Hamiltonian as the sum of several components, e.g.

$$H = H^\circ + H' \tag{2.2}$$

where the solutions of the wave equation

$$H_i^\circ \psi_i^\circ = E_i^\circ \psi_i^\circ \tag{2.3}$$

are known exactly. The energy levels and wave functions associated with equation (2.3) may be written as column matrices which in general will be infinite.

$$\mathbf{E}^\circ = \begin{bmatrix} E_1^\circ \\ \vdots \\ E_i^\circ \\ \vdots \\ E_n^\circ \end{bmatrix}; \qquad \boldsymbol{\psi}^\circ = \begin{bmatrix} \psi_1^\circ \\ \vdots \\ \psi_i^\circ \\ \vdots \\ \psi_n^\circ \end{bmatrix} \tag{2.4}$$

The wave functions of the operator H, equation (2.2), may be expressed as linear combinations of these wave functions,

$$\psi_j = \sum_{i=1}^{n} C_{ji} \psi_i^\circ \tag{2.5}$$

where the ψ_i° are referred to as the set of basis functions for the problem. Equation (2.5) may be written in matrix notation as,

$$\boldsymbol{\psi} = \mathbf{C} \boldsymbol{\psi}^\circ \tag{2.6}$$

where,

$$\boldsymbol{\psi} = \begin{bmatrix} \psi_1 \\ \vdots \\ \psi_i \\ \vdots \\ \psi_n \end{bmatrix} \qquad \text{and} \qquad \mathbf{C} = \begin{bmatrix} C_{11} \cdots C_{1i} \cdots C_{1n} \\ \vdots \quad\quad \vdots \quad\quad \vdots \\ C_{i1} \cdots C_{ii} \cdots C_{in} \\ \vdots \quad\quad \vdots \quad\quad \vdots \\ C_{n1} \cdots C_{ni} \cdots C_{nn} \end{bmatrix}$$

The wave equation associated with equation (2.2) may be solved by converting the Hamiltonian energy matrix (2.7) into a diagonal matrix (2.8) consisting of the allowed energy levels of the system,

$$\mathbf{H} = \begin{bmatrix} H_{11} \cdots H_{1i} \cdots H_{1n} \\ \vdots \qquad \vdots \qquad \vdots \\ H_{i1} \cdots H_{ii} \cdots H_{in} \\ \vdots \qquad \vdots \qquad \vdots \\ H_{n1} \cdots H_{ni} \cdots H_{nn} \end{bmatrix} \qquad (2.7)$$

$$\mathbf{E} = \begin{bmatrix} E_1 \\ \quad E_2 \\ \qquad \ddots \qquad\quad 0 \\ \qquad\qquad \ddots \\ \quad 0 \qquad\qquad \ddots \\ \qquad\qquad\qquad\quad E_n \end{bmatrix} \qquad (2.8)$$

The individual elements of equation (2.7) are given by

$$H_{ij} = \int \psi_i^{\circ *} H \psi_j^{\circ} \, d\tau$$

where * denotes the complex conjugate and the integration is carried out over all the variables which define the problem.

The notation

$$H_{ij} = \langle m_i n_i \ldots | H | m_j n_j \ldots \rangle$$

where $m, n \ldots$ are the quantum numbers required to define a state of the system represented by the Hamiltonian operator of equation (2.3) is frequently used to represent matrix elements.

The conversion of equation (2.7) into (2.8) may be achieved by a transformation

$$\mathbf{E} = \mathbf{C}^+ \mathbf{H} \mathbf{C} \qquad (2.9)$$

where $\mathbf{C}$ is the matrix of the coefficients in the expansion of the wave functions and $\mathbf{C}^+$ is its transpose. The operation implied by equation (2.9) is usually carried out using a computer and it is necessary when the matrix (2.7) is infinite to limit the order of the matrices. This is done by truncating the set of basis functions at some point which is generally a compromise between computing time and the accuracy desired. The energy levels may also be obtained by expansion and solution of a secular determinant of the form given by equation (2.10),

$$
\begin{vmatrix}
H_{11}-E & \cdots H_{1i} & \cdots H_{1n} \\
\vdots & \vdots & \vdots \\
H_{i1} & \cdots H_{ii}-E & \cdots H_{in} \\
\vdots & \vdots & \vdots \\
H_{n1} & \cdots H_{ni} & \cdots H_{nn}-E
\end{vmatrix} = 0 \qquad (2.10)
$$

Finding the energy levels by equation (2.9) rather than (2.10) has the advantage that it also gives the wave functions through the matrix $\mathbf{C}$.

The average values of any operator F may be obtained as the diagonal elements of the matrix $\mathbf{F}'$,

$$\mathbf{F}' = \mathbf{C}^{+}\mathbf{F}\mathbf{C} \qquad (2.11)$$

where the elements of the matrix $\mathbf{F}$ are defined by,

$$F_{ij} = \int \psi_i^{\circ*} F \psi_j^{\circ} \, d\tau$$

When F is the dipole moment operator μ, the diagonal elements of $\mathbf{F}'$ correspond to average values of the dipole moment in the different states while the off-diagonal elements may be related to the strengths of the spectroscopic transitions between the different states.

If H' is small compared to H° then the energy levels and wave functions may be obtained by perturbation theory.[2] The Hamiltonian matrix (2.7) may be written as,

$$
\begin{bmatrix}
E_1^{\circ} + H'_{11} \cdots & H'_{1i} & \cdots & H'_{1n} \\
\vdots & \vdots & & \vdots \\
H'_{i1} & \cdots E_i^{\circ} + H'_{ii} \cdots & H'_{in} \\
\vdots & \vdots & & \vdots \\
H'_{n1} & \cdots & H'_{ni} & \cdots E_n^{\circ} + H'_{nn}
\end{bmatrix} \qquad (2.12)
$$

where the E_i^0 are the solutions of equation (2.3) and,

$$H'_{ij} = \int \psi_i^{\circ*} H' \psi_j^{\circ} \, d\tau$$

The energy levels of the Hamiltonian operator (2.2) correct to the zeroeth order are simply the E_i°, while those correct to the first order are the diagonal elements of the matrix (2.12). The energy levels correct to the second order are given by,

$$E_i = E_i^\circ + H'_{ii} + \sum_{j \neq i} \frac{(H'_{ij})^2}{(E_i^\circ - E_j^\circ)}$$

The wave functions correct to the first order are

$$\psi_i = \psi_i^\circ + \sum_{j \neq i} \frac{H'_{ij}\psi_j^\circ}{(E_i^\circ - E_j^\circ)}$$

Sometimes, by the correct choice of basis functions or by ordering the basis functions in a certain way, it is possible to reduce the Hamiltonian matrix to the form shown in equation (2.13) in which there are no elements connecting the submatrices $H_1, H_2, H_3, \ldots$.

$$H = \begin{bmatrix} H_1 & & \\ & H_2 & \\ & & H_3 \end{bmatrix} \qquad (2.13)$$

In such a case the energy levels of the system are found by diagonalizing each of the sub-matrices separately. This factorization of the Hamiltonian matrix is a consequence of the symmetry of the Hamiltonian operator and the allowed wave functions for the system, and some examples will be found later in this chapter. It is very useful when a large set of basis functions has to be used since the computing time required to diagonalize matrices increases very rapidly with their size.

The factorization of the Hamiltonian matrix discussed in the previous paragraph is completely rigorous. Sometimes when this is not feasible it is possible to achieve an approximate factorization using a technique known as the Van Vleck transformation.[3] This may be applied in a situation where the Hamiltonian matrix has the form shown in equation (2.14). A particular sub-matrix, e.g. H_1, contains a set of relatively closely spaced energy levels while the energy separation between different sub-matrices is relatively large.

$$H = \begin{bmatrix} H_1 & \lambda & \lambda \\ \lambda & H_2 & \lambda \\ \lambda & \lambda & H_3 \end{bmatrix} \qquad (2.14)$$

There are small matrix elements represented by λ connecting the different sub-matrices. A transformation of the type given by equation (2.9) is applied to the Hamiltonian matrix so that the matrix elements connecting different blocks are reduced to the order λ^2 as shown in equation (2.15).

$$\bar{\mathbf{H}} = \begin{bmatrix} \bar{\mathbf{H}}_1 & \lambda^2 & \lambda^2 \\ \lambda^2 & \bar{\mathbf{H}}_2 & \lambda^2 \\ \lambda^2 & \lambda^2 & \bar{\mathbf{H}}_3 \end{bmatrix} \tag{2.15}$$

The elements of the sub-matrices $\bar{\mathbf{H}}_i$ of equation (2.15) are modified by the transformation when compared to those of $\mathbf{H}_i$ of equation (2.14) The energy levels of the system are obtained by neglecting the matrix elements connecting the different sub-matrices of equations (2.15) and diagonalizing each sub-matrix $\bar{\mathbf{H}}_i$ separately. The Van Vleck transformation is frequently used in the study of large amplitude vibrations. In such cases the sub-matrices represent different vibrational states and the more closely spaced levels within a particular sub-matrix are pure rotational energy levels.

2.3 The Molecular Hamiltonian Operator and Energy Levels

There are considerable differences in the way theoreticians and experimentalists approach a problem such as internal rotation, although the ultimate aims of gaining insight into the nature of intramolecular forces are identical. This difference in approach occurs mainly because it is not possible to solve the Schrödinger equation (2.1) exactly for molecular systems and therefore apparently different models and approximations must be adopted. Molecular spectroscopy shows that to a high level of accuracy the total energy of an isolated molecule can usually be expressed as the sum of its electronic, vibrational, rotational and translational energies,[4]

$$E_{\text{mol}} = E_{\text{elec}} + E_{\text{vib}} + E_{\text{rot}} + E_{\text{trans}} \tag{2.16}$$

Some additional terms must be added to equation (2.16) to take into account the magnetic effects due to electron and nuclear spins and external electric and magnetic fields. The separation of the energy into its separate components is not completely rigorous and in some cases it is necessary to add terms to equation (2.16) which allow for the interaction between the electronic and vibrational motions or between the vibration and rotational motions. It will be seen that vibration–rotation interactions often occur in molecules which have large amplitude vibrations. To the same degree of approximation that equation (2.16) is valid the molecular Hamiltonian operator may be written as a sum of separate operators,

$$H_{mol} = H_{elec} + H_{vib} + H_{rot} + H_{trans} \tag{2.17}$$

and the molecular wave function as a product of separate wave functions,

$$\psi_{mol} = \psi_{elec} \times \psi_{vib} \times \psi_{rot} \times \psi_{trans} \tag{2.18}$$

The separate equations derived from equations (2.16) to (2.18) are,

$$H_{elec}\psi_{elec} = E_{elec}\psi_{elec} \tag{2.19}$$

$$H_{vib}\psi_{vib} = E_{vib}\psi_{vib} \tag{2.20}$$

$$H_{rot}\psi_{rot} = E_{rot}\psi_{rot} \tag{2.21}$$

$$H_{trans}\psi_{trans} = E_{trans}\psi_{trans} \tag{2.22}$$

The equation involving the translational energy of the molecule (2.22) is of no interest as far as large amplitude vibrations are concerned. Equation (2.19) is the starting point for theoretical investigations and its application to molecules with large amplitude vibrations will be discussed more fully in Chapter 4. The vibrational and rotational equations (2.20) and (2.21) are the respective starting points for infrared and Raman or microwave studies and will occupy later sections of this chapter.

The Hamiltonian operator for a molecule can be written in terms of operators representing the kinetic energies of the nuclei (T_n), electrons (T_e) and the coulombic interactions (V) between the different particles,

$$H_{mol} = T_n + T_e + V_{nn} + V_{ne} + V_{ee} \tag{2.23}$$

In order to convert equation (2.23) into (2.17) it is necessary to make some changes of coordinate systems and approximations. Equation (2.23) applies to a coordinate system fixed in space as shown in Fig. 2.1(a). A transformation to a coordinate system fixed at the centre of mass of the molecule (Fig. 2.1(b)) makes it possible to separate the translational energy of the molecule. The Born–Oppenheimer approximation makes it possible to separate the electronic and nuclear motions. Because of their small mass electrons in a molecule are moving very rapidly compared to the nuclei. This means that as far as the electrons are concerned the nuclei may be considered to be fixed in space at any give instant of time. The Hamiltonian operator for the electronic motion is thus,

$$H_{elec} = T_e + V_{nn} + V_{ne} + V_{ee}$$

As far as the vibrational and rotational motions are concerned molecules may be regarded as consisting of point atomic masses. This is another consequence of the small electronic mass and is also due to atoms retaining to a

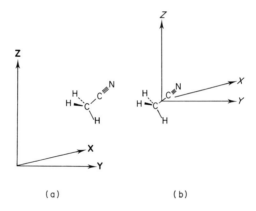

(a)   (b)

Fig. 2.1 Showing (a) location of the atoms of a molecule (e.g. CH_3CN) relative to a space fixed axis system, (b) use of an axis system located at the centre of mass which leads to the separation of the translational energy.

large extent spherical electronic charge distributions even when chemically bound in a molecule. The separation of the vibrational and rotational motions involves choosing the molecule fixed axes in such a way that no angular momentum is generated as a result of the molecular vibrations (Eckart condition). This is only rigorously possible if the molecular vibrations are of truly small amplitude. The breakdown of the separation of the rotational and vibrational motions is discussed more fully in Section 2.9.

2.4 Electronic States and Potential Energy Surfaces

In the majority of molecules the electronic states are widely separated from each other in terms of energy and for the ground electronic state the Born–Oppenheimer approximation is particularly good. In some molecules, especially those with unpaired electrons and orbital angular momentum, there are very low excited electronic states and this may lead to very large interactions between the electronic and vibrational motions. In linear molecules such vibronic interactions are known as Renner–Teller effects while in nonlinear molecules they are known as Jahn–Teller effects.[5] Both of these effects can give rise to large amplitude vibrations of a special nature which will not be discussed further in this book. All of the large amplitude motions described later will not depend in any way on the breakdown of the Born–Oppenheimer approximation.

The electronic energy is a complicated function of the internal coordinates used to describe the displacements of the nuclei from their equilibrium positions. In Chapter 1 it was mentioned that the electronic or vibrational potential energy could be represented as a surface in a multidimensional space. Molecules in which internal rotation or inversion occur are characterized by having potential energy surfaces which have a number of adjacent minima separated by relatively low maxima. Other molecules have potential energy surfaces with single but very ill-defined minima. In such cases it is possible to distort the molecule appreciably from its equilibrium position before large restoring forces are encountered. This gives rise to large amplitude bending or wagging vibrations.

In the experimental study of large amplitude vibrations it is seldom possible to obtain sufficient information to be able to work backwards and reconstruct the potential energy surface in the way this can be done for the potential energy curve for a diatomic molecule. In most cases it is necessary to consider that the large amplitude motion only involves a change in one internal coordinate, consequently only a section through the potential energy surface is examined. In internal rotation, for example, this does not necessarily describe the energetically most economical path to bring about the interconversion of two rotamers. There is some experimental evidence to show that a simple one dimensional approach to large amplitude vibrations is not strictly correct, for example, the value of the barrier to inversion in ammonia or the barrier to internal rotation in methanol is found to depend on the vibrational state of the molecule. Also non equivalent rotamers of a molecule cannot always be interconverted by a simple rotation about a bond. In spite of such shortcomings the one dimensional approach to large amplitude motions is amazingly successful. Important exceptions are found in some ring molecules and some molecules containing a linear or nearly linear chain of atoms where it is necessary to use two internal coordinates to describe the large amplitude ring puckering or bending vibrations.

2.5 Molecular Vibrations

2.5.1 Introduction

Having discussed the origin of the vibrational potential energy surface we are now in a position to discuss the nature of molecular vibrations. In the following chapters the reader will find that large amplitude vibrations are treated in a way which separates them from the remaining small amplitude vibrations of the particular molecule under consideration. For example, the treatment of internal rotation in acetaldehyde presupposes the molecule to be composed of rigid methyl and acetyl groups connected by a rigid carbon–carbon bond. Thus, the torsional angle about the carbon–carbon bond represents the only variable parameter, and the internal motion reduces to that of a one dimensional Hamiltonian operator and potential function. In order that we can explore the validity of the total neglect of all molecular vibrations other than the torsional mode (see later sections of this chapter), it is necessary at this juncture to summarize some of the more important points about small amplitude molecular vibrations.

2.5.2 Small amplitude molecular vibrations and classical mechanics

For a molecule consisting of N atoms, the total number of internal degrees of freedom (excluding translational and rotational motions) is $3N - 5$ for a linear system and $3N - 6$ for the general case. The kinetic and potential energies associated with the movements of the constituent atoms may thus be expressed in terms of $3N - 6$ internal coordinates. These internal coordinates measure the deviations of structural parameters such as bond lengths and bond angles from their equilibrium values. Four commonly used classes of internal coordinates are given in Fig. 2.2.

The use of "n" internal coordinates ($n = 3N - 6$) rather than space fixed cartesian coordinates ensures that the molecular vibrations may be treated separately from the overall translational and rotational motions of the

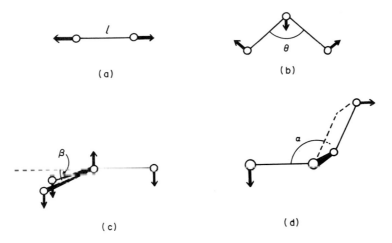

Fig. 2.2   Four common types of internal coordinates, (a) bond stretching, δl, (b) bond angle bending, $\delta \theta$, (c) out of plane bending, $\delta \beta$, (d) torsion about a bond, $\delta \alpha$, where δl represents the change in l etc.

molecule.[6] However, the separation of the vibrational and overall rotational motions is not complete and will be dealt with in later sections of this chapter. One important simplifying assumption which is made for small amplitude vibrations is that atoms move in straight lines.

Information about potential energy surfaces in terms of harmonic and anharmonic force constants[7] usually has to be derived from vibration spectra, since the determination of molecular potential energy surfaces from molecular orbital calculations is feasible only for a relatively small number of simple molecules. The potential energy may be approximated by a Taylor series expansion about the equilibrium structure in terms of the internal coordinates (q_i),

$$V = V_0 + \sum_{i=1}^{n} \left(\frac{\partial V}{\partial q_i}\right)_e q_i + \tfrac{1}{2} \sum_{i,j=1}^{n} \left(\frac{\partial^2 V}{\partial q_i \partial q_j}\right)_e q_i q_j$$

$$+ \tfrac{1}{6} \sum_{i,j,k=1}^{n} \left(\frac{\partial^3 V}{\partial q_i \partial q_j \partial q_k}\right)_e q_i q_j q_k + \cdots \qquad (2.24)$$

Equation (2.24) may be simplified by setting the potential energy V_0 of the equilibrium structure equal to zero, and all of the partial derivatives $(\partial V / \partial q_i)_e$ vanish because they express the condition for the potential energy to be a minimum at the equilibrium configuration. The harmonic force constants are defined as,

$$F_{ij} = \left(\frac{\partial^2 V}{\partial q_i \partial q_j}\right)_e$$

and the cubic (anharmonic) constants as,

$$F_{ijk} = \left(\frac{\partial^3 V}{\partial q_i \partial q_j \partial q_k}\right)_e$$

Since the cubic (and higher) terms are so much smaller than the quadratic ones we will simplify our treatment further by neglecting all terms other than the harmonic terms. The potential energy thus becomes,

$$V = \tfrac{1}{2} \sum_{i,j=1}^{n} F_{ij} q_i q_j \tag{2.25}$$

In terms of the matrix of the force constants the potential energy may be written as,

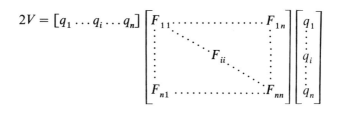

or

$$2V = \mathbf{q}^+ \mathbf{F} \mathbf{q} \tag{2.26}$$

The vibrational kinetic energy (T) of the molecule may similarly be expressed in terms of a matrix which is given the symbol $\mathbf{G}$. The kinetic energy is given by,

$$T = \tfrac{1}{2} \sum_{i,j=1}^{n} G_{ij} p_i p_j \tag{2.27}$$

where the p_i terms are the momenta conjugate to the coordinates q_i. The G_{ij} are functions of the geometric parameters of the molecule and the atomic masses,

$$G_{ij} = \sum_{k=1}^{n} \frac{B_{ik} B_{jk}}{m_k}$$

where B_{ik}, B_{jk} represent coefficients which relate the internal coordinates to the cartesian displacement coordinates of the moving atoms. The **G** matrix represents the inverse of the matrix of the reduced masses, and the matrix elements G_{ij} may be obtained from molecular parameters using simple rules (see ref. 1, Further Reading). Using matrix notation, equation (2.27) may be written,

$$2T = [p_1 \cdots p_i \cdots p_n] \begin{bmatrix} G_{11} \cdots \cdots \cdots \cdots \cdots G_{1n} \\ \vdots \quad \ddots \qquad \qquad \vdots \\ \vdots \qquad \ddots G_{ll} \qquad \vdots \\ \vdots \qquad \qquad \ddots \quad \vdots \\ G_{n1} \cdots \cdots \cdots \cdots \cdots G_{nn} \end{bmatrix} \begin{bmatrix} p_1 \\ \vdots \\ p_l \\ \vdots \\ p_n \end{bmatrix}$$

or

$$2T = \mathbf{p}^+ \mathbf{G} \mathbf{p} \tag{2.28}$$

The kinetic energy of the system can also be expressed in terms of the time derivatives of the coordinates $(\dot{q}_i)$,

$$2T = \dot{\mathbf{q}}^+ \mathbf{G}^{-1} \dot{\mathbf{q}} \tag{2.29}$$

where $\mathbf{G}^{-1}$ is the inverse of the matrix **G**.

In order to derive a particularly convenient set of equations for the modes and the frequencies of the molecular vibrations, a new set of coordinates can be defined $(Q_1, \ldots, Q_n)$ in terms of the internal coordinates,

$$
\begin{aligned}
Q_1 &= R_{11}q_1 + \ldots R_{1i}q_i + \ldots R_{1n}q_n \\
&\vdots \qquad \vdots \qquad \qquad \vdots \qquad \qquad \vdots \\
Q_i &= R_{i1}q_1 + \ldots R_{ii}q_i + \ldots R_{in}q_n \\
&\vdots \qquad \vdots \qquad \qquad \vdots \qquad \qquad \vdots \\
Q_n &= R_{n1}q_1 + \ldots R_{ni}q_i + \ldots R_{nn}q_n
\end{aligned}
\tag{2.30}
$$

where R_{ij} represents the appropriate normalized transformation coefficients (i.e. $\sum_j R_{ij}^2 = 1$). The choice of these new coordinates ensures that the matrix expressions for the potential and kinetic energies (V and T) are diagonal (i.e. the coefficients of all the cross terms in equations (2.25) and (2.27) are zero). The respective expressions for the two energy terms then become,

$$2V = \sum_{i=1}^{n} \lambda_i Q_i^2$$

$$\tag{2.31}$$

$$2T = \sum_{i=1}^{n} \dot{Q}_i^2$$

These equations represent the energies of "n" independent harmonic oscillators whose vibrational frequencies (v_i) are related to the coefficients λ_i as follows,

$$\lambda_i = 4\pi^2 v_i^2 \tag{2.32}$$

The coordinates Q_i are called the normal coordinates and any arbitrary vibrational motion of the molecule may be expressed as a linear combination of them,

$$\mathbf{q} = \mathbf{LQ} \tag{2.33}$$

(where $\mathbf{L} = \mathbf{R}^+$; see equation (2.30)).

The vibrations associated with the normal coordinates are called the normal vibrations and during the execution of a normal mode, all the atoms vibrate in phase with each other with the same frequency. However, the amplitudes of vibration of different atoms vary for each normal vibration, and this forms the basis of the concept of localized group vibrations in molecules.

We may rewrite the two expressions in equation (2.31) in matrix notation as follows,

$$2V = \mathbf{Q}^+\lambda\mathbf{Q}$$
$$2T = \dot{\mathbf{Q}}^+\mathbf{E}\dot{\mathbf{Q}} \tag{2.34}$$

Here $\mathbf{E}$ is the unit diagonal matrix, and λ is a diagonal matrix consisting of the λ_i terms.

From equations (2.26), (2.29), (2.33) and (2.34) we can derive the following relationships,

$$\mathbf{L}^+\mathbf{FL} = \lambda \tag{2.35}$$

and

$$\mathbf{L}^+\mathbf{G}^{-1}\mathbf{L} = \mathbf{E} \tag{2.36}$$

and hence deduce the convenient secular equation (determinant),

$$|\mathbf{GF} - \lambda\mathbf{E}| = 0 \tag{2.37}$$

The solutions of this secular equation correspond to the frequencies of the normal vibrations.

If the molecule has elements of symmetry the vibrational problem can be simplified by capitalizing on this. Symmetry coordinates can be utilized to factorize the secular equation into a number of smaller blocks. These co-ordinates, which are linear combinations of the internal coordinates, are chosen so as to be symmetric or antisymmetric with respect to the symmetry operations which characterize the molecule (see Section 2.5.4).

2.5.3 Small amplitude molecular vibrations and quantum mechanics

In terms of the normal coordinates, the Hamiltonian of a molecular system can be written as,

$$H = \frac{1}{2}\left[\sum_{i=1}^{n} \dot{Q}_i^2 + \sum_{i=1}^{n} \lambda_i Q_i^2\right] \tag{2.38}$$

and the Schrödinger wave equation becomes,

$$-\frac{h^2}{8\pi^2}\sum_{i=1}^{n}\frac{\partial^2 \psi}{\partial Q_i^2} + \frac{1}{2}\sum_{i=1}^{n}\lambda_i Q_i^2 \psi = E\psi \tag{2.39}$$

The use of normal coordinates ensures that in quantum mechanics, just as in classical mechanics, the total vibrational problem separates into indi-vidual parts, one for each vibrational mode. Thus equation (2.39) represents the wave equation for the sum of "n" independent harmonic oscillators. The wave function is the product of "n" harmonic oscillator wave functions,

$$\psi = \psi_1 \psi_2 \ldots \psi_i \ldots \psi_n$$

and the energy is given by,

$$E = E_1 + E_2 + \ldots E_i + \ldots E_n$$

where

$$E_i = (v_i + \tfrac{1}{2})h\nu_i$$

v_i being the ith vibrational quantum number. Thus the quantum mechanical energy levels may be obtained very easily from the classical solutions to the secular equation (2.37).

2.5.4 The separation of molecular vibrations[8]

As mentioned before it is sometimes possible, by virtue of the molecular symmetry, to simplify the secular equation into smaller and more easily manageable equations. When a symmetry coordinate s_i is the only one belonging to a particular symmetry class then factorization results in a simple one dimensional block. For example, in a simple non linear XY_2 molecular system, the symmetry coordinates may be chosen as linear combinations of internal coordinates as shown in Fig. 2.3. For the XY_2

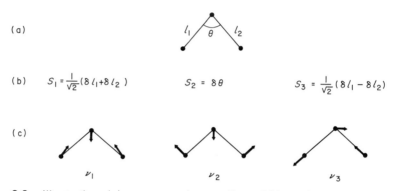

Fig. 2.3   Illustrating (a) a symmetric non-linear XY_2 molecule with internal coordinates, δl_1, δl_2 and $\delta\theta$, (b) symmetry coordinates, and (c) the normal modes of vibration.

molecule, s_3 has different symmetry properties to s_1 and s_2. (It is anti-symmetric with respect to reflection in the plane bisecting the angle θ.) For a one dimensional block (such as s_3) the frequency may be found from the product of the F and G matrix elements,

$$F_{ii}G_{ii} = \lambda_i$$
$$= 4\pi^2 v_i^2 \tag{2.40}$$

Here F_{ii} is given by,

$$F_{ii} = F_{l_1 l_1} - F_{l_1 l_2}$$

where $F_{l_1 l_1}$ is the bond stretching force constant and $F_{l_1 l_2}$ is the interaction constant between the bond stretches and is concerned with the contribution to the potential energy arising from the products of the two internal bond stretching coordinates.

In instances where symmetry plays an important role the factorization of the secular equation is exact. However, an approximate separation can be achieved when one frequency in a particular block differs considerably from the others. Consideration of a two dimensional block shows this,

$$\begin{vmatrix} (\mathbf{GF})_{11} - \lambda & (\mathbf{GF})_{12} \\ (\mathbf{GF})_{21} & (\mathbf{GF})_{22} - \lambda \end{vmatrix} = 0$$

where

$$(\mathbf{GF})_{11} = G_{11}F_{11} + G_{12}F_{21}, \qquad (\mathbf{GF})_{12} = G_{11}F_{12} + G_{12}F_{22}$$

$$(\mathbf{GF})_{21} = G_{21}F_{11} + G_{22}F_{21}, \qquad (\mathbf{GF})_{22} = G_{21}F_{12} + G_{22}F_{22}$$

The high and low frequency vibrations become separable when $(\mathbf{GF})_{12}$ and $(\mathbf{GF})_{21}$ are small compared to $(\mathbf{GF})_{11}$ and $(\mathbf{GF})_{22}$. Under these conditions the two vibrational frequencies are given closely by,

$$\lambda_1 = (\mathbf{GF})_{11} \quad \text{and} \quad \lambda_2 = (\mathbf{GF})_{22}$$

Good conditions for this type of factorization are realized, for example, in hydrocarbons where the C—H stretching modes are all around 3000 cm^{-1} and all other vibrations are below 1700 cm^{-1}.

This separation of high and low frequencies is one of the main reasons why the treatment of large amplitude vibrations using simple one dimensional potential functions is so successful. Large amplitude vibrations are generally characterized by very low vibrational frequencies and anharmonic potential functions. In view of the frequency difference it would appear plausible to consider large amplitude vibrations separately from others. The anharmonic cross terms in the potential energy function are unlikely to prevent this unless there is a close degeneracy of levels which are connected by one of these terms. In such cases both levels will be pushed away from each other as shown in Fig. 2.4.

This type of effect is unlikely to be met with in the lower energy levels of the large amplitude vibration and does not often affect the determination of the potential function. Although the height and general shape of the potential function hindering a large amplitude motion is often drastically altered if another vibration in the molecule is excited, it is usually found that the energy levels associated with the large amplitude vibration may still be adequately fitted using a one dimensional potential function.

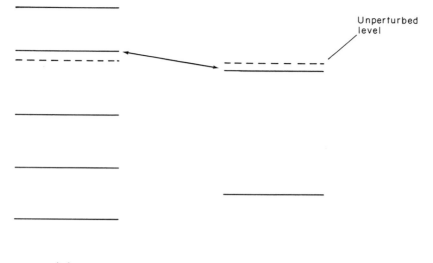

Fig. 2.4   (a) Manifold of relatively closely spaced levels of a large amplitude vibration. (b) Manifold of widely spaced levels of a small amplitude vibration. The figure shows how vibrational energy levels can be perturbed by anharmonic cross terms in the potential function. The positions of the unperturbed levels are indicated by dotted line.

2.6 The Mathematical Representation of Potential Energy Curves for Large Amplitude Vibrations

In previous sections of this chapter it has been shown that often, to a very high degree of approximation, large amplitude vibrations may be separated from the other vibrations of the molecule and treated as one dimensional problems. The energy levels are calculated by solving one dimensional wave equations with appropriate reduced masses and potential functions. The potential functions used to describe large amplitude motions may be divided into two types:

(a) periodic functions
(b) non periodic functions.

The first type are used to describe internal rotation while the second type are used for inversion and some types of ring vibrations.

2.6.1 Periodic potential functions

Any continuous periodic function of a single variable such as that shown in Fig. 2.5(a) may be represented by a Fourier series,

$$F(\alpha) = b_0 + \sum_{n=1}^{\infty} b_n \cos n\alpha + \sum_{n=1}^{\infty} c_n \sin n\alpha$$

The curve shown in Fig. 2.5(b) has the shape expected for the potential energy function describing the internal rotation in an asymmetric molecule such as a derivative of methanol of the type CXY_2—OH.[9] Most of the molecules with internal rotation studied so far have symmetrical potential functions in the sense that $V(\alpha) = V(\pi + \alpha)$. Since $\cos n\alpha = \cos n(\pi + \alpha)$ while $\sin n\alpha = -\sin n(\pi + \alpha)$ the potential functions for these molecules may be expressed as a cosine Fourier series,

$$V(\alpha) = \sum_{n=0}^{\infty} V'_n \cos n\alpha \qquad (2.41)$$

By a suitable choice of origin, Fig. 2.5(c), this may be written in the form,

$$V(\alpha) = \sum_{n=1}^{\infty} \frac{V_n}{2}(1 - \cos n\alpha) \qquad (2.42)$$

Apart from the additive constant V'_0 of equation (2.41) and $\sum_{n=1}^{\infty} V_n/2$ of equation (2.42) these functions are equivalent, therefore $V'_n = V_n/2$.

The problem of determining a potential function for an internal rotation motion thus amounts to evaluating a sufficient number of the coefficients in the Fourier series expansion to define the function with a reasonable degree of accuracy. If the potential function has high symmetry then many of the coefficients are automatically zero, e.g. for a methyl group only V_3, V_6, V_9, etc., are non-vanishing. In such cases V_6 and higher coefficients have generally been found to be very small compared to V_3 and the series is said to converge very rapidly. When the potential function has lower symmetry a greater number of coefficients is necessary to describe it and often it is only possible to determine a few of these from the available experimental data. Usually the series is truncated at a certain point by assuming that the higher coefficients are zero.

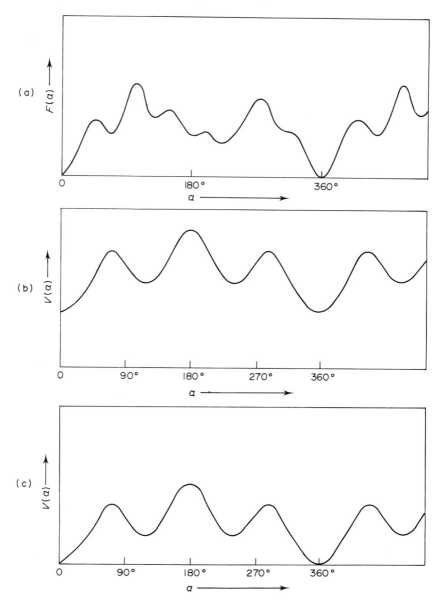

Fig. 2.5   Periodic curves that may be represented by (a) a general Fourier series, (b) equation (2.41), (c) equation (2.42).

2.6.2 Non-periodic potential functions

In Chapters 7 and 8 non-periodic one dimensional potential functions will be used to describe the inversion in amine molecules and ring puckering

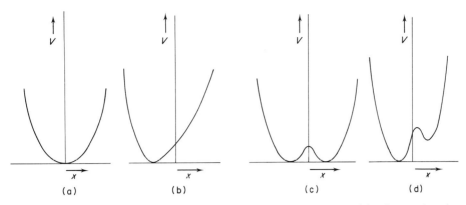

Fig. 2.6   Some non-periodic potential functions encountered in the study of large amplitude vibrations, (a) symmetrical single minimum, (b) asymmetrical single minimum, (c) symmetrical double minimum, (d) asymmetrical double minimum.

vibrations and these may have the shape of one of the functions shown in Fig. 2.6. The simplest way to represent functions of this type is to use power series expansions in a coordinate x. The expansions for symmetrical functions, Fig. 2.6(a) and (c), contain only even powers of x,

$$V(x) = \sum_{n=1}^{\infty} a_{2n} x^{2n} \tag{2.43}$$

while expansions for functions with no symmetry, Fig. 2.6(b) and (d), contain both even and odd powers of x,

$$V(x) = \sum_{n=2}^{\infty} a_n x^n \tag{2.44}$$

In equation (2.43) the coordinate x measures the deviation of the system from the planar or symmetrical configuration. The absence of a linear term in x from equation (2.44) means that the origin of the coordinate system does not necessarily correspond, for example, to the planar configuration of an inverting amine group or to a planar skeleton of the ring atoms in a puckering vibration. A number of other types of function have been used to represent such potentials, but some of these were chosen more to facilitate solution of the Schrödinger equation than to give accurate values for the potential energy.

As in the case of internal rotation, it is generally only possible to determine a limited number of the coefficients from the available data and equations (2.43) and (2.44) are rarely taken beyond terms in x^4. These truncated functions will not give accurate values for the potential energy for large values of x.

2.7 Calculation of the Energy Levels Associated with One-dimensional Potential Functions

In this section the methods outlined in Section 2.2 will be applied to simple examples of large amplitude vibrations involving periodic and non-periodic potential functions. A number of assumptions will be made so that the kinetic energy part of the Hamiltonian operator may be written in a very simple form. The first example could, with a little modification, be used to give the torsional energy levels for a molecule such as nitrobenzene while the second example could be used to give the energy levels for the inversion of an amine molecule. It is important to notice how the relative spacings of the energy levels change from the case where there is no barrier hindering the internal motion to the case where it is prevented by a very high barrier.

2.7.1 Internal rotation hindered by a symmetrical twofold barrier

The simple Hamiltonian operator,

$$H = -\frac{d^2}{d\alpha^2} + \frac{V_2}{2}(1 - \cos 2\alpha) \qquad (2.45)$$

will be considered and this may be written as

$$H = H° + H'$$

where

$$H° = -\frac{d^2}{d\alpha^2}$$

and

$$H' = \frac{V_2}{2}(1 - \cos 2\alpha)$$

$H°$ is the one dimensional Hamiltonian operator for free rotation and the wave functions derived from the Schrödinger equation,

$$H°\psi = E\psi \qquad (2.46)$$

may be used as a set of basis functions to obtain the energy levels and wave functions associated with equation (2.45). The reader may verify by substitution into equation (2.46) that suitable wave functions and the corresponding levels of $H°$ are,

$$\psi_n = \frac{1}{\sqrt{2\pi}} e^{\pm in\alpha}$$

$$E_n = n^2 \quad \text{where} \quad n = 0, \pm 1, \pm 2 \ldots, \pm n.$$

Table 2.1   Matrix elements associated with the Hamiltonian operator:

$$H = -\frac{d^2}{d\alpha^2} + \frac{V_l}{2}(1 - \cos l\alpha)$$

$$e^{\pm in\alpha} \text{ basis}$$

$$\left\langle n \left| \frac{V_l}{2} \right| n \right\rangle = \frac{V_l}{2}$$

$$\left\langle n \left| \frac{V_l}{2} \right| m \right\rangle = 0 \, (n \neq m)$$

$$\langle n | \cos l\alpha | n + l \rangle = \langle n + l | \cos l\alpha | n \rangle - \tfrac{1}{2}$$

$$\langle n | \cos l\alpha | m \rangle = 0 \, (n \neq m \pm l)$$

$$\left\langle n \left| \frac{d^2}{d\alpha^2} \right| n \right\rangle = -n^2$$

$$\left\langle n \left| \frac{d^2}{d\alpha^2} \right| m \right\rangle = 0 \, (n \neq m)$$

$$\sin n\alpha \text{ and } \cos n\alpha \text{ basis}$$

$\sin n\alpha$	$\cos n\alpha$

$$\left\langle n \left| \frac{V_l}{2} \right| n \right\rangle = \frac{V_l}{2} \qquad\qquad \left\langle n \left| \frac{V_1}{2} \right| n \right\rangle = \frac{V_l}{2}$$

$$\left\langle n \left| \frac{V_l}{2} \right| m \right\rangle = 0 \, (n \neq m) \qquad\qquad \left\langle n \left| \frac{V_l}{2} \right| m \right\rangle = 0 \, (n \neq m)$$

$$\left\langle n \left| \frac{d^2}{d\alpha^2} \right| n \right\rangle = -n^2 \qquad\qquad \left\langle n \left| \frac{d^2}{d\alpha^2} \right| n \right\rangle = -n^2$$

$$\left\langle n \left| \frac{d^2}{d\alpha^2} \right| m \right\rangle = 0 \, (n \neq m) \qquad\qquad \left\langle n \left| \frac{d^2}{d\alpha^2} \right| m \right\rangle = 0 \, (n \neq m)$$

$$\langle n | \cos l\alpha | n + l \rangle = \langle n + l | \cos l\alpha | n \rangle = \tfrac{1}{2} \qquad \langle 0 | \cos l\alpha | l \rangle = \langle l | \cos l\alpha | 0 \rangle = \frac{1}{\sqrt{2}}$$

$$\langle n | \cos l\alpha | m \rangle = -\tfrac{1}{2} \quad (l = m + n) \qquad\qquad \langle n | \cos l\alpha | m \rangle = \tfrac{1}{2} \quad (l = m \pm n)$$

$$\langle n | \cos l\alpha | m \rangle = 0 \quad \begin{pmatrix} n \neq m \pm l \\ n \neq l - m \end{pmatrix} \qquad \langle n | \cos l\alpha | m \rangle = 0 \quad \begin{pmatrix} n \neq m \pm l \\ n \neq l - m \end{pmatrix}$$

The matrix elements associated with the Hamiltonian operator of equation (2.45) are given in Table 2.1, and the Hamiltonian matrix has the form,

$$\mathbf{H} = \begin{array}{c} \begin{array}{ccccccc} n \quad -3 & -2 & -1 & 0 & 1 & 2 & 3 \end{array} \\ \begin{array}{c} -3 \\ -2 \\ -1 \\ 0 \\ 1 \\ 2 \\ 3 \end{array} \left[\begin{array}{ccccccc} 9+\dfrac{V_2}{2} & 0 & -\dfrac{V_2}{4} & 0 & 0 & 0 & 0 \\[2mm] 0 & 4+\dfrac{V_2}{2} & 0 & -\dfrac{V_2}{4} & 0 & 0 & 0 \\[2mm] -\dfrac{V_2}{4} & 0 & 1+\dfrac{V_2}{2} & 0 & -\dfrac{V_2}{4} & 0 & 0 \\[2mm] 0 & -\dfrac{V_2}{4} & 0 & \dfrac{V_2}{2} & 0 & -\dfrac{V_2}{4} & 0 \\[2mm] 0 & 0 & -\dfrac{V_2}{4} & 0 & 1+\dfrac{V_2}{2} & 0 & -\dfrac{V_2}{4} \\[2mm] 0 & 0 & 0 & -\dfrac{V_2}{4} & 0 & 4+\dfrac{V_2}{2} & 0 \\[2mm] 0 & 0 & 0 & 0 & -\dfrac{V_2}{4} & 0 & 9+\dfrac{V_2}{2} \end{array} \right] \end{array} \quad (2.47)$$

The functions,

$$\psi_0^{(1)} = \frac{1}{\sqrt{2\pi}}$$

$$\psi_n^{(1)} = \frac{1}{\sqrt{\pi}} \cos n\alpha, \quad \text{where} \quad n = 1, 2, 3, \ldots, n.$$

$$\psi_n^{(2)} = \frac{1}{\sqrt{\pi}} \sin n\alpha, \quad \text{where} \quad n = 1, 2, 3, \ldots, n.$$

are also acceptable wave functions for H° and if the Hamiltonian matrix is set up using these functions all of the matrix elements (Table 2.1) of the type $\int \psi_n^{(1)} H \psi_m^{(2)} \, d\alpha$ are zero.[10] The Hamiltonian matrix may then be represented by two submatrices,

$$\mathbf{H} = \begin{bmatrix} \mathbf{H}^1 & 0 \\ 0 & \mathbf{H}^2 \end{bmatrix}$$

and the energy levels may be obtained by diagonalizing $\mathbf{H}^1$ and $\mathbf{H}^2$ separately.

The matrices $\mathbf{H}^1$ and $\mathbf{H}^2$ may be constructed using the elements associated with $\cos n\alpha$ and $\sin n\alpha$ basis functions respectively as given in Table 2.1 and are shown below.

$$
\mathbf{H}^1 =
\begin{bmatrix}
\dfrac{V_2}{2} & 0 & -\dfrac{\sqrt{2}}{4}V_2 & 0 & 0 & 0 \\[2ex]
0 & 1+\dfrac{V_2}{4} & 0 & -\dfrac{V_2}{4} & 0 & 0 \\[2ex]
-\dfrac{\sqrt{2}}{4}V_2 & 0 & 4+\dfrac{V_2}{2} & 0 & -\dfrac{V_2}{4} & 0 \\[2ex]
0 & -\dfrac{V_2}{4} & 0 & 9+\dfrac{V_2}{2} & 0 & -\dfrac{V_2}{4} \\[2ex]
0 & 0 & -\dfrac{V_2}{4} & 0 & 16+\dfrac{V_2}{2} & 0
\end{bmatrix}
\tag{2.48}
$$

$$
\mathbf{H}^2 =
\begin{bmatrix}
1+\dfrac{3V_2}{4} & 0 & -\dfrac{V_2}{4} & 0 & 0 & 0 \\[2ex]
0 & 4+\dfrac{V_2}{2} & 0 & -\dfrac{V_2}{4} & 0 & 0 \\[2ex]
-\dfrac{V_2}{4} & 0 & 9+\dfrac{V_2}{2} & 0 & -\dfrac{V_2}{4} & 0 \\[2ex]
0 & -\dfrac{V_2}{4} & 0 & 16+\dfrac{V_2}{2} & 0 & -\dfrac{V_2}{4} \\[2ex]
0 & 0 & -\dfrac{V_2}{4} & 0 & 25+\dfrac{V_2}{2} & 0
\end{bmatrix}
\tag{2.49}
$$

The results of diagonalizing $\mathbf{H}^1$ and $\mathbf{H}^2$ for different values of V_2 are shown in Fig. 2.7 and Table 2.2. For low values of the barrier height the energy levels are similar to those for free internal rotation but in the limit of a very high barrier they become equivalent to those of a harmonic oscillator with force constant $k = 2V_2$. The energy levels for intermediate barriers may be labelled using either the free rotation or high barrier quantum numbers.

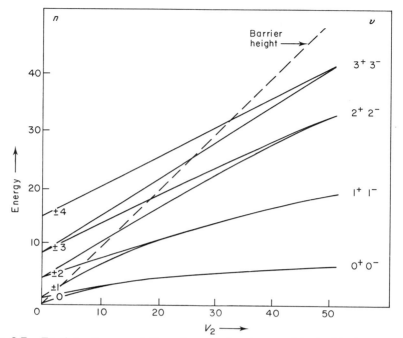

Fig. 2.7   To show the correlation of energy levels associated with the potential function $V(\alpha) = V_2/2 \ (1 - \cos 2\alpha)$. The units of energy (including V_2) are arbitrary.

Table 2.2   Some of the energy levels, in arbitrary units, associated with the Hamiltonian operator

$$H = -\frac{d^2}{d\alpha^2} + \frac{V_2}{2} \ (1 - \cos 2\alpha)$$

for the different values of V_2

V_2 \ n	0	+1	-1	+2	-2	+3	-3
0·0	0·000	1·000	1·000	4·000	4·000	9·000	9·000
0·1	0·050	1·025	1·075	4·050	4·050	9·050	9·050
0·5	0·242	1·123	1·373	4·249	4·257	9·251	9·251
1·0	0·469	1·242	1·742	4·495	4·526	9·504	9·504
5·0	1·819	2·083	4·524	6·371	7·050	11·569	11·629
10·0	2·847	2·924	7·496	8·492	10·613	14·186	14·612
50·0	6·811	6·811	19·862	19·867	31·618	31·718	41·420
100·0	9·743	9·743	28·685	28·685	46·478	46·479	62·964
100·0[a]	10·000	10·000	30·000	30·000	50·000	50·000	70·000
v	0^+	0^-	1^+	1^-	2^+	2^-	3^+

[a] Calculated using the simple harmonic oscillator approximation, equation (2.40) with the force constant $F_{ii} = 2V_2$ and $G_{ii} = 8\pi^2$.

2.7.2 The energy levels associated with a simple non-periodic potential function

The simple potential function

$$V(x) = x^4 - ax^2 \qquad (2.50)$$

will be used to illustrate some features of the energy levels and wave functions of non-periodic potential functions. When the constant a of equation (2.50) is negative the potential has a single minimum and is similar to that shown in Fig. 2.6(a). When a is positive the potential is a double minimum function like that shown in Fig. 2.6(c). The reader may verify by differentiation of equation (2.50) with respect to x and imposing the condition for minima that these occur at $x = \pm\sqrt{a/2}$. The barrier height is $a^2/4$. In order to derive the energy levels associated with the above potential function it is convenient to write the Hamiltonian operator as,

$$H = v_0(P^2 + X^4 - AX^2) \qquad (2.51)$$

where v_0 is a scale factor and P and X are a dimensionless momentum operator and coordinate respectively. The coordinate X and the constant A of equation (2.51) are related to x and a of equation (2.50) through v_0 and the reduced mass for the vibration.

If simple harmonic oscillator wave functions are used as a basis set to obtain the energy levels (E_i) associated with equation (2.51) it is convenient to choose v_0 as the frequency of simple harmonic oscillator. The matrix elements of P^2, X^2 and X^4 in this basis are given in Table 2.3,[11] and the Hamiltonian matrix (apart from the multiplicative constant v_0) is,

n	0	1	2	3	4
0	$4 - A$	0	$\sqrt{2}(5 - A)$	0	$2\sqrt{6}$
1	0	$18 - 3A$	0	$\sqrt{6}(9 - A)$	0
2	$\sqrt{2}(5 - A)$	0	$44 - 5A$	0	$\sqrt{12}(13 - A)$
3	0	$\sqrt{6}(9 - A)$	0	$82 - 7A$	0
4	$2\sqrt{6}$	0	$\sqrt{12}(13 - A)$	0	$132 - 9A$

Since there are no elements connecting odd and even quantum numbers the Hamiltonian matrix factors into two sub-matrices containing only even and

Table 2.3   Matrix elements associated with the Hamiltonian operator

$$H = P^2 + X^4 - AX^2$$

$\langle n|P^2|n\rangle = 2n + 1$

$\langle n|P^2|n + 2\rangle = \langle n + 2|P^2|n\rangle = -\left[(n + 1)(n + 2)\right]^{\frac{1}{2}}$

$\langle n|P^2|m\rangle = 0 \quad \left(\begin{matrix} n \neq m \\ n \neq m \pm 2 \end{matrix}\right)$

$\langle n|X^2|n\rangle = 2n + 1$

$\langle n|X^2|n + 2\rangle = \langle n + 2|X^2|n\rangle = \left[(n + 1)(n + 2)\right]^{\frac{1}{2}}$

$\langle n|X^4|n\rangle = 3(2n^2 + 2n + 1)$

$\langle n|X^4|n + 2\rangle = \langle n + 2|X^4|n\rangle = (4n + 6)\left[(n + 1)(n + 2)\right]^{\frac{1}{2}}$

$\langle n|X^4|n + 4\rangle = \langle n + 4|X^4|n\rangle = \left[(n + 1)(n + 2)(n + 3)(n + 4)\right]^{\frac{1}{2}}$

$\langle n|X^4|m\rangle = 0 \quad \left(\begin{matrix} n \neq m \\ n \neq m \pm 2 \\ n \neq m \pm 4 \end{matrix}\right)$

---

odd quantum numbers respectively. This factorization is a consequence of equation (2.51) being symmetric with respect to the operation $X \rightarrow -X$ while the harmonic oscillator wave functions are symmetric for even values of the quantum number n and antisymmetric for odd values of n. A Hamiltonian of the type (2.51) is referred to as a reduced Hamiltonian operator and X as a reduced coordinate. A further example of a reduced Hamiltonian operator may be found in Chapter 7.

The first few energy levels obtained by diagonalizing the Hamiltonian matrices associated with equation (2.51) for different values of A are summarized in the form of a correlation diagram in Fig. 2.8 and Table 2.4. In the high barrier limit the energy level pattern resembles that of a simple harmonic oscillator except that the levels are doubly degenerate. The members of a pair of these levels are distinguished by a $+$ or $-$ suffix which denotes the symmetry of the wave function with respect to the operation $X \rightarrow -X$. In the region of intermediate barriers the energy levels may be labelled by either the limiting high or low barrier schemes. In order to obtain the energy levels with sufficient accuracy it is frequently necessary to diagonalize large matrices. Table 2.5 gives the average values of $X^2(\langle X^2 \rangle)$ for different values of the barrier height. For the intermediate barrier case the irregular variation of $\langle X^2 \rangle$ should be noted since it has important consequences in connection with the pure rotational spectra of molecules which have a large amplitude vibration describable by a potential function of this type (Section 2.8.4).

Table 2.4 Energy levels, in units of ν_0, associated with the Hamiltonian operator

$$H = P^2 + X^4 - AX^2$$

in the harmonic oscillator basis

A	Barrier height $(A^2/4)$	$n=0$	$n=1$	$n=2$	$n=3$	$n=4$	$n=5$	$n=6$
0·0	0·0000	2·6719	9·5746	18·7872	29·3429	40·9722	53·5173	66·8476
0·1	0·0025	2·6166	9·4333	18·5916	29·0976	40·6869	53·1852	66·4761
0·5	0·0625	2·4380	8·9067	17·8500	28·1570	39·5662	51·8973	65·0308
1·0	0·2500	2·3091	8·3322	17·0143	27·0721	38·2564	50·3784	63·3153
5·0	6·2500	4·4082	6·6398	14·1374	21·9872	31·3918	41·8412	53·2073
10·0	25·0000	8·5236	13·1860	21·8407	24·1573	32·0814	40·3492	49·3415
50·0	625·0000	19·9193	19·9193	59·4293	59·4293	98·4334	98·4334	136·9156
50·0[a]	625·0000	20·0000	20·0000	60·0000	60·0000	100·0000	100·0000	140·0000
v		0⁺	0⁻	1⁺	1⁻	2⁺	2⁻	3⁺

[a] calculated using the simple harmonic oscillator approximation equation (2.40). The force constant $F_{ii} = d^2V/dX^2$, evaluated at $X = \sqrt{A/2}$, is $4A\nu_0$, and $G_{ii} = 32\pi^2\nu_0$.

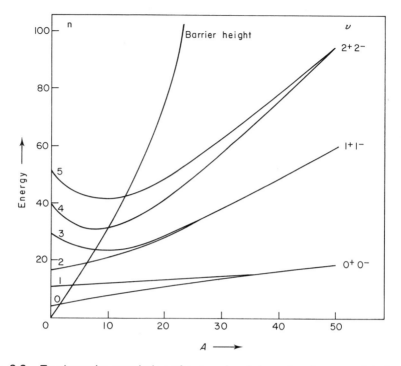

Fig. 2.8   To show the correlation of energy levels in units of v_0 associated with the Hamiltonian operator

$$H = P^2 + X^4 - AX^2$$

Table 2.5   Average values of $X^2 (\langle X^2 \rangle)$ associated with the potential function $V(x) = X^4 - AX^2$

A	$n = 0$	$n = 1$	$n = 2$	$n = 3$	$n = 4$
0·0	0·57	1·43	1·98	2·47	2·92
0·1	0·58	1·44	1·99	2·48	2·93
0·5	0·61	1·49	2·02	2·52	2·97
1·0	0·65	1·56	2·07	2·57	3·01
5·0	1·50	2·40	2·27	2·98	3·41
10·0	4·53	3·03	2·48	3·56	4·23
50·0	24·80	24·80	24·39	24·39	23·53

2.8 Molecular Rotation

2.8.1 Rotation in classical mechanics

In classical mechanics the kinetic energy of a rigid rotating body is,

$$T = \tfrac{1}{2}(I_{xx}\omega_x^2 + I_{yy}\omega_y^2 + I_{zz}\omega_z^2 - 2I_{xy}\omega_x\omega_y - 2I_{xz}\omega_x\omega_z - 2I_{yz}\omega_y\omega_z) \quad (2.52)$$

where $\omega_x, \omega_y, \omega_z$ are the components of the angular velocity of the body about the axes x, y, z of a coordinate system fixed in the body, I_{xx}, I_{yy}, and I_{zz} are the moments of inertia and I_{xy}, I_{xz} and I_{yz} are the products of inertia. If the axes are chosen to pass through the centre of mass of the body the moments and products of inertia are given by equations of the form,

$$I_{xx} = \sum_{i=1}^{n} m_i(y_i^2 + z_i^2) \qquad (2.53)$$

$$I_{xy} = \sum_{i=1}^{n} - m_i x_i y_i \qquad (2.54)$$

where the summations are extended over all of the particles which make up the body. Expressions for the remaining moments and products of inertia may be obtained by permuting the x, y, z subscripts in equations (2.53) and (2.54).

Equation (2.52) may be expressed conveniently in matrix form as,

$$2T = \omega^+ I \omega$$

where the matrix I is known as the inertial tensor and is,

$$I = \begin{bmatrix} I_{xx} & I_{xy} & I_{xz} \\ I_{xy} & I_{yy} & I_{yz} \\ I_{xz} & I_{yz} & I_{zz} \end{bmatrix}$$

and ω^+ and ω are respectively row and column matrices with components ω_x, ω_y and ω_z.

It is always possible to choose an axis system such that the three products of inertia are zero and this is referred to as the principal inertial axis system. By convention the principal axes are labelled a, b, c where the a axis is associated with the smallest and the c axis with the largest moment of inertia. The principal axis system is related to an arbitrary axis system by the transformation which diagonalizes the inertial tensor in the arbitrary axis frame,

$$
\begin{bmatrix} I_a & 0 & 0 \\ 0 & I_b & 0 \\ 0 & 0 & I_c \end{bmatrix} = \begin{bmatrix} C_{11} & C_{21} & C_{31} \\ C_{12} & C_{22} & C_{32} \\ C_{13} & C_{23} & C_{33} \end{bmatrix} \begin{bmatrix} I_{xx} & I_{xy} & I_{xz} \\ I_{xy} & I_{yy} & I_{yz} \\ I_{xz} & I_{yz} & I_{zz} \end{bmatrix} \begin{bmatrix} C_{11} & C_{12} & C_{13} \\ C_{21} & C_{22} & C_{23} \\ C_{31} & C_{32} & C_{33} \end{bmatrix}
$$

and coordinates in the principal axis system are related to those in the original axis system by,

$$
\begin{bmatrix} a \\ b \\ c \end{bmatrix} = \begin{bmatrix} C_{11} & C_{12} & C_{13} \\ C_{21} & C_{22} & C_{23} \\ C_{31} & C_{32} & C_{33} \end{bmatrix} \begin{bmatrix} x \\ y \\ z \end{bmatrix}
$$

In terms of the principal moments of inertia the kinetic energy of rotation is,

$$T = \tfrac{1}{2}(I_a\omega_a^2 + I_b\omega_b^2 + I_c\omega_c^2) \tag{2.55}$$

Equation (2.55) may be expressed in terms of angular momenta using relationships of the type $P_a = I_a\omega_a$ and the kinetic energy becomes,

$$T = \frac{1}{2}\left(\frac{P_a^2}{I_a} + \frac{P_b^2}{I_b} + \frac{P_c^2}{I_c}\right) \tag{2.56}$$

In classical mechanics the kinetic energy and the total angular momentum of a freely rotating body are constants of the motion and in quantum mechanics these quantities have discrete and constant values. Equation (2.56), after the proper transformation of the angular momenta to the corresponding quantum mechanical operators, provides the basis for discussing the rotational energy levels of molecules.

2.8.2 Inertial classification of molecules

Molecules may be divided into four classes depending on the relationships between their principal moments of inertia, and the different classes have characteristic patterns of rotational energy levels.

(a) Linear molecules. For a molecular model consisting of point masses a linear molecule has $I_a = 0$ and $I_b = I_c$ and consequently $P_a = 0$ and $P_b = P_c$.

(b) Symmetric top molecules. Molecules with a threefold or higher axis of symmetry have two equal principal moments of inertia and belong to this class. Prolate symmetric tops have $I_b = I_c$ and a mass distribution which resembles a cigar, e.g. methyl chloride CH_3Cl. Oblate symmetric tops have $I_a = I_b$ and their mass distribution is discoid, e.g. benzene C_6H_6. It is important to note that an isotopically substituted molecule such as CH_2DCl is not a symmetric top.

(c) Spherical top molecules. This type has all three moments of inertia equal and only molecules with very high symmetry, e.g. methane CH_4 belong to this class. In dealing with linear molecules, symmetric and spherical top molecules it is customary to designate the equal moments of inertia by I.

(d) Asymmetric top molecules. The majority of molecules having all three moments of inertia different belong to this class.

If an asymmetric top molecule has a plane of symmetry then one of the principal axes will be perpendicular to this plane and the other two will lie in it. For the important case of a rigid planar molecule the c inertial axis is perpendicular to the molecular plane and from the equations for the moments of inertia (2.53) it is easy to see that,

$$I_c = I_a + I_b$$

For real molecules this relationship is replaced by,

$$\Delta = I_c - I_a - I_b$$

where Δ, the inertial defect, is generally a small positive quantity. The inertial defect is the consequence of molecular vibration, and is important in discussions of molecular planarity.[12]

2.8.3 Rigid rotor energy levels

In quantum mechanics the angular momentum of a freely rotating body is a constant,

$$P = \sqrt{J(J+1)}\ \frac{h}{2\pi} \tag{2.57}$$

where the rotational quantum number J can take the values 0, 1, 2, . . . The rotational energy levels of rigid linear molecules, symmetric top and spherical top molecules may be derived very simply from equations (2.56)

c

and (2.57) because of the relationships between the moments of inertia and the components of the angular momentum.[13]

(a) *Linear molecules*. The Hamiltonian operator for a linear molecule may be written as,

$$H = \frac{P^2}{2I} \tag{2.58}$$

by making the substitutions $P_a = 0$, $P_b = P_c$ and $I_b = I_c = I$ into equation (2.56). Squaring equation (2.57) and substituting the values for P^2 into equation (2.56) gives,

$$E_J = J(J + 1)\frac{h^2}{8\pi^2 I}$$

The energy levels are usually expressed in frequency units (MHz) or in wave numbers (cm^{-1}) and the quantity $h/8\pi^2 I$ (or $h/8\pi^2 cI$) is called the rotational constant and is given the symbol B. The rotational energy levels of a rigid linear molecule may therefore be expressed as,

$$E_J = BJ(J + 1)$$

and in the absence of an external electric or magnetic field are specified by the single quantum number J and have a degeneracy of $2J + 1$. In the presence of an external field the component of the angular momentum in the field direction is restricted to values (in units of $h/2\pi$)

$$M_J = 0, \pm 1, \pm 2, \pm \ldots J$$

In an electric field the energy depends on M_J^2 and so a twofold degeneracy, except for $M_J = 0$, remains for each level.

(b) *Symmetric top molecules*. The Hamiltonian operator for a prolate symmetric top molecule is,

$$H = \frac{1}{2}\left(\frac{P_a^2}{I_a} + \frac{P^2 - P_a^2}{I}\right)$$

and if the rotational constants A and B are used this may be written as,

$$H = AP_a^2 + (P^2 - P_a^2)B$$

In classical mechanics the component of the angular momentum along the symmetry axis of a symmetric top is also a constant of the motion. In quantum mechanics for a prolate symmetric top.

$$P_a = K \frac{h}{2\pi} \text{ where } K = 0, \pm 1, \pm 2, \dots, \pm J$$

and the energy levels are given by.

$$E_{JK} = BJ(J + 1) + (A - B)K^2$$

The arrangement of the energy levels of a prolate symmetric top is shown in Fig. 2.9(a). For an oblate symmetric top,

$$P_c = K \frac{h}{2\pi}$$

and the rotational energy levels are given by,

$$E_{JK} = BJ(J + 1) + (C - B)K^2$$

The arrangement of the energy levels for an oblate symmetric top is shown in Fig. 2.9(b). Compared to a linear molecule, the energy levels of a symmetric top with K not equal to zero have an additional degeneracy factor of two. In an electric field the energy levels with K different from zero split into $2J + 1$ components while that for K equal to zero splits into $J + 1$ components.

(c) *Spherical top molecules.* The Hamiltonian operator for this type of molecule is simply,

$$H = BP^2$$

and this gives the same formula as a linear molecule for the energy levels. The energy levels have no dependence on the quantum number K and in the absence of an external field each level has a degeneracy of $(2J + 1)^2$.

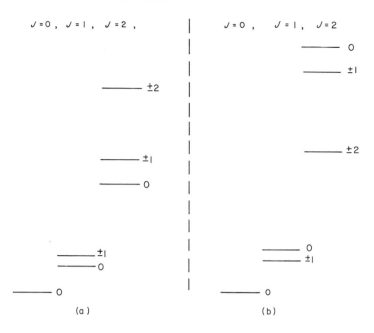

Fig. 2.9   Part of the energy level manifold of (a) a prolate symmetric top, (b) an oblate symmetric top. Each level is labelled by the appropriate K quantum number.

(d) *Asymmetric top molecules*. Except for a few levels of low rotational quantum number the energy levels of asymmetric top molecules cannot be expressed in a simple form. The wave function for a particular asymmetric top energy level may be expressed as a linear combination of the symmetric top wave functions of the same rotational quantum number J. The Hamiltonian matrix may be set up using these basis functions and diagonalized to give the energy levels. By making full use of the symmetry of the Hamiltonian operator and the symmetric top wave functions it is possible to divide the Hamiltonian matrix into four sub-matrices for each value of J.

In asymmetric tops there are $2J + 1$ different energy levels for each value of J and these are usually labelled using the notation J_{K_{-1}, K_1}, where K_{-1} and K_1 are respectively the K levels of the limiting prolate and oblate symmetric tops with which the level correlates. A correlation diagram for the lowest J levels is shown in Fig. 2.10. The K_{-1} and K_1 subscripts are labels and not quantum numbers and sometimes an alternative scheme, J_τ where $\tau = K_{-1} - K_1$, is used to label the energy levels. The degree of asymmetry

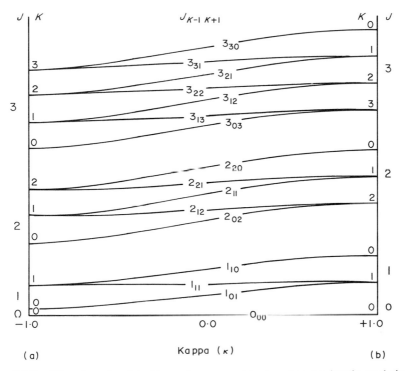

Fig. 2.10 Showing the labelling of asymmetric top energy levels and their correlation with the energy levels of the limiting (a) prolate and (b) oblate symmetric tops.

of the rotor may be expressed, for example, using Ray's asymmetry parameter,

$$\kappa = \frac{2B - A - C}{A - C}$$

and for a prolate symmetric top $\kappa = -1$ and for an oblate symmetric top $\kappa = +1$.

2.8.4 Effective rigid rotor energy levels and rotational constants

The rotational energy levels of the ground and excited states of molecules are generally found to follow rigid rotor theory quite accurately but it is necessary to have a slightly different set of effective or empirical rotational constants for each vibrational state. This is readily understood since the rotational constants are related to the inverse of the moments of inertia averaged over the vibrational state of the molecule. The effective rotational constants also contain contributions which arise from the separation of the vibrational and rotational motions of the molecule. For molecules in which there are no large amplitude vibrations the effective rotational constants may be expressed in the form,

$$B_v = B_e + \sum_i (v_i + \tfrac{1}{2})\alpha_i + \sum_i (v_i + \tfrac{1}{2})^2 \beta_i$$

where B_e is the equilibrium rotational constant and the v_i are the vibrational quantum numbers that characterize the state. The α_i and β_i are constants related to the quadratic, cubic and quartic terms of the vibrational potential function. For simple molecules such as HCN and SO_2 the α_i and β_i may be used to help determine the anharmonic potential constants.[14]

When a low frequency large amplitude vibration occurs it is frequently possible to assign the rotational spectra of a number of vibrational states which correspond to the excitation of successive quanta of the large amplitude vibration. In such cases the variation of the rotational constants with vibrational state may provide useful information about the nature of the potential function of the large amplitude vibration and if the contributions from the interaction between this vibration and the molecular rotation are removed from the rotational constants, the latter may be interpreted in a geometrical manner. They may be expressed as series in terms of the average values of the powers of the coordinate used to describe the large amplitude vibration. When the potential function is symmetrical these series take the form,[15]

$$B_v = B_e + \beta_1 \langle x^2 \rangle + \beta_2 \langle x^4 \rangle + \cdots$$

For a single minimum potential function the rotational constants usually vary smoothly with increasing vibrational quantum number while for a double minimum potential function with a moderate or low barrier there is often a pronounced zig–zag variation of the rotational constants as shown in Fig. 2.11. The rotational constants are so sensitive to the nature of the potential function that in oxetane $((CH_2)_3O)$ it has been possible to show that there is a barrier of less than $0 \cdot 2 \, \text{kJ mol}^{-1}$ preventing the inversion of the ring. This is considerably less than the zero point energy of the ring puckering vibration and it is almost a philosophical point as to whether or not the molecule has a planar ring.

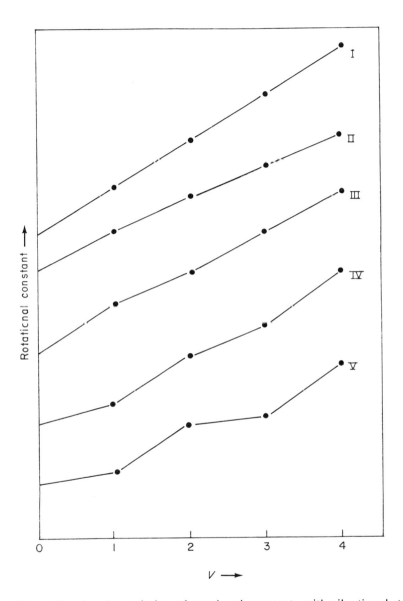

Fig. 2.11   Showing the variation of rotational constants with vibrational state for: a planar molecule with (I) a harmonic vibration, (II) an anharmonic vibration, and a non-planar molecule with (III) a very low barrier, (IV) a medium barrier and (V) a high barrier.

2.8.5 Centrifugal distortion

Molecular energy levels deviate to a small extent from the effective rigid rotor model discussed in the previous section. These deviations increase with increasing rotational energy and are due to centrifugal distortion.[16] An appreciation of the factors involved may be obtained from a classical treatment of a diatomic molecule consisting of two masses held together by harmonic forces, and rotating about a single axis. This may be regarded as a semi-rigid model since, although the bond length changes as a result of the centrifugal distortion, the molecule is not considered to be vibrating. As the molecule rotates there is a centrifugal force tending to stretch the bond given by,

$$f_c = - \frac{dE_{rot}}{dr}$$

$$= - \frac{P^2}{2} \frac{d(1/I)}{dr}$$

and this is balanced by a restoring force of $k\Delta r$, where k is the force constant of the bond and Δr is the change in bond length brought about by the centrifugal distortion. Equating the two forces gives,

$$\Delta r = - \frac{P^2}{2k} \frac{d(1/I)}{dr} \tag{2.58}$$

The change in the bond length produces a decrease in the rotational energy of,

$$\Delta E_{rot} = \frac{P^2}{2} \frac{d(1/I)}{dr} \Delta r$$

and substituting for Δr from equation (2.58) gives,

$$\Delta E_{rot} = - \frac{P^4}{4k} \left(\frac{d(1/I)}{dr} \right)^2$$

There is also an increase in the potential energy of,

$$\Delta E_{pot} = \tfrac{1}{2} k \Delta r^2$$

$$= \frac{P^4}{8k} \left(\frac{d(1/I)}{dr} \right)^2$$

and the total change in energy due to centrifugal distortion is,

$$\Delta E_{cd} = \Delta E_{rot} + \Delta E_{pot}$$

$$= -\frac{P^4}{8k}\left(\frac{d(1/I)}{dr}\right)^2 \tag{2.59}$$

If the quantum mechanical value of P^4 is substituted into equation (2.59) the centrifugal distortion energy for a diatomic molecule is,

$$E_{cd} = -\frac{J^2(J+1)^2}{8k}\left(\frac{d(1/I)}{dr}\right)^2 \hbar^4$$

$$= -J^2(J+1)^2 D_J \tag{2.60}$$

where D_J is the centrifugal distortion constant.

Equation (2.60) can be generalized for polyatomic molecules to take into account angular momenta about all three principal axes and more than one force constant and derivative of the inverse moment of inertia with respect to internal coordinate. Centrifugal distortion constants have been used to provide information about the quadratic force constants of simple linear, symmetric top and asymmetric top molecules. In molecules with large amplitude vibrations anomalous centrifugal distortion effects may be expected because of the weak restoring forces and their anharmonic nature.

2.9 Vibration Rotation Interactions

The small differences in the effective rotational constants of different vibrational states and centrifugal distortion are two aspects of the interaction between the molecular vibrational and rotational motions. Another type of vibration rotation interaction is found in molecules with two vibrational states which are degenerate or very nearly equal in energy and which satisfy certain symmetry requirements. The physical origin of these effects is the coupling of the angular momentum generated as a result of the molecular vibrations with the angular momentum due to the overall rotation of the molecule and they are generally referred to as Coriolis interactions. These interactions can lead to marked departures from rigid rotor behaviour for some of the rotational energy levels of the two states while the remaining rotational levels and the other vibrational states are not significantly affected. In linear and symmetric top molecules there are degenerate vibrations and

Coriolis interactions lead to a splitting of some of the pairs of rotation vibration energy levels which otherwise would be degenerate. This phenomenon is referred to as l-type doubling and is frequently encountered in the excited states of the low frequency bending vibrations of these molecules. It will be recalled from the discussion of Section 2.7 that periodic potential functions or symmetrical double minimum potential functions with a high barrier can give rise to degenerate or nearly degenerate pairs of energy levels. Coriolis interactions are often prominent features of the microwave spectra of molecules with these types of potential functions and can be analysed to give the separation in energy between the two vibrational or torsional states. It will be seen later that such information is particularly useful in helping to determine potential functions.

The manner in which the rotational energy levels in an asymmetric rotor are affected by a Coriolis type of interaction may be understood by considering the Hamiltonian operator and wave functions. For simplicity the molecule will be assumed to be rigid except for a single large amplitude motion which gives rise to a pair of almost degenerate levels (v_1, v_2) and vibrational angular momentum about the a inertial axis. The Hamiltonian operator is,[17]

$$H = \underbrace{AP_a^2 + BP_b^2 + CP_c^2}_{H_1} + \underbrace{(F/2)p^2 + V(x)}_{H_2} - \underbrace{F'pP_a}_{H_3} \tag{2.61}$$

where F and F' are parameters related to the molecular geometry. The first term (H_1) represents the rotational energy of the molecule, the second term (H_2) the vibrational energy and the third term (H_3) the vibration–rotation interaction. If the last term in equation (2.61) were zero the energy levels would be the sum of the vibrational energy and that of a rigid motor with the effective rotational constants of the vibrational state being considered. The total wave functions would be products of the vibrational wave functions and rigid rotor wave functions,

e.g. $\psi_{v_1 J_1 \tau_1} = \psi_{v_1} \psi_{J_1 \tau_1}; \quad \psi_{v_2 J_2 \tau_2} = \psi_{v_2} \psi_{J_2 \tau_2}$

and the rotational wave functions would be linear combinations of appropriate symmetric top wave functions.

The part of the Hamiltonian operator representing the vibration–rotation interaction is the product of the operator p which depends only on the vibrational wave functions and the operator P_a which depends only on the rotational wave functions. If the matrix element $\langle v_1|p|v_2\rangle$ is not zero then the rotational energy levels of the two states cannot be obtained by treating them as two independent effective rigid rotors. The matrix elements $\langle J_1\tau_1|P_a|J_2\tau_2\rangle$ are zero unless J_1 and J_2 are equal and the Hamiltonian matrix therefore factorizes into sub-matrices for each value of J of the form,

$$
\left[
\begin{array}{c|c}
A_1 P_a^2 + B_1 P_b^2 + C_1 P_c^2 + E_{v_1} & F_p' P_a \\
\hline
F_p' P_a & A_2 P_a^2 + B_2 P_b^2 + C_2 P_c^2 + E_{v_2}
\end{array}
\right]
\quad (2.62)
$$

The diagonal blocks of equation (2.62) are simply matrices representing the sum of the vibrational and rigid rotor energies and the off diagonal block introduces matrix elements which connect rotational levels in states v_1 and v_2. If symmetric top wave functions are used as a basis set for the rigid rotor wave functions then the Hamiltonian matrix for $J = 1$ is given by expression (2.63).

$$
\begin{vmatrix}
0{\cdot}5(B_1 + C_1) + A_1 + E_{v_1} & 0 & -0{\cdot}5(B_1 - C_1) \\
0 & B_1 + C_1 + E_{v_1} & 0 \\
-0{\cdot}5(B_1 - C_1) & 0 & 0{\cdot}5(B_1 + C_1) + A_1 + E_{v_1} \\
-G & 0 & 0 \\
0 & 0 & 0 \\
0 & 0 & G
\end{vmatrix}
$$

$$
\begin{matrix}
-G & 0 & 0 \\
0 & 0 & 0 \\
0 & 0 & G \\
0{\cdot}5(B_2 + C_2) + A_2 + E_{v_2} & 0 & -0{\cdot}5(B_2 - C_2) \\
0 & B_2 + C_2 + E_{v_2} & 0 \\
-0{\cdot}5(B_2 - C_2) & 0 & 0{\cdot}5(B_2 + C_2) + A_2 + E_{v_2}
\end{matrix}
$$

$$(2.63)$$

where $G = F'\langle v_1|p|v_2\rangle$.

Table 2.6   Comparison of the values of the $J = 1$ rotational energy levels for the $v = 1$ and $v = 2$ vibrational states. (a) in the absence of vibration–rotation interactions. (b) by diagonalization of matrix (2.63). (c) by second-order perturbation theory

		$v = 1$	$v = 2$
(a)	1_{01}	$B_1 + C_1 + E_{v_1} = E_1$	$B_2 + C_2 + E_{v_2} = E_2$
	1_{11}	$A_1 + C_1 + E_{v_1} = E_3$	$A_2 + C_2 + E_{v_2} = E_4$
	1_{10}	$A_1 + B_1 + E_{v_1} = E_5$	$A_2 + B_2 + E_{v_2} = E_6$
(b)	1_{01}	E_1	E_2
	1_{11}	$0.5[E_3 + E_6 - ((E_3 + E_6)^2 - 4(E_3 E_6 - G^2))^{\frac{1}{2}}]$	$0.5[E_4 + E_5 \pm ((E_4 + E_5)^2 - 4(E_4 E_5 - G^2))^{\frac{1}{2}}]$
	1_{10}	$0.5[E_4 + E_5 \mp ((E_4 + E_5)^2 - 4(E_4 E_5 - G^2))^{\frac{1}{2}}]$	$0.5[E_3 + E_6 + ((E_3 + E_6)^2 - 4(E_3 E_6 - G^2))^{\frac{1}{2}}]$
(c)	1_{01}	E_1	E_2
	1_{11}	$E_3 + G^2/(E_3 - E_6)$	$E_4 + G^2/(E_4 - E_5)$
	1_{10}	$E_5 + G^2/(E_5 - E_4)$	$E_6 + G^2/(E_6 - E_3)$

The energy levels obtained by diagonalization of the matrix (2.63) are compared to those in the absence of the vibration–rotation interaction contributions in Table 2.6. The vibration–rotation interaction may also be treated by second-order perturbation theory and the non-zero matrix elements H_{ij} of the matrix (2.62) connecting the rotational levels of the v_1 and v_2 states are shown schematically in Figure 2.12. The corrections for the different energy levels are also given in Table 2.6.

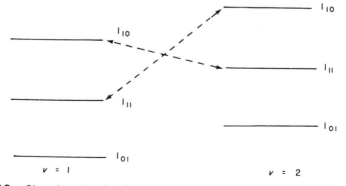

Fig. 2.12   Showing the $J = 1$ rotational energy levels of the v_1 and v_2 vibrational states connected by the operator P_a.

Further Reading

1. E. B. Wilson Jr., J. C. Decius and P. C. Cross. "Molecular Vibrations". McGraw-Hill, New York 1955.
   The classic text on advanced analysis of molecular vibrations.
2. H. C. Allen Jr, and P. C. Cross. "Molecular Vib-Rotors". Wiley, Chichester and New York, 1963.
   An advanced treatise on rotation and vibration–rotation spectra of molecules.
3. M. D. Harmony. "Introduction to Molecular Energies and Spectra". Holt, Rinehart and Winston, New York, 1972.
   An excellent introductory text dealing with molecular energy levels and the various types of possible spectra between them.
4. J. M. Anderson. "Mathematics for Quantum Chemistry". Benjamin, New York, 1966.
   A good introductory book on mathematical methods used in quantum chemistry.

References

1. P. W. Atkins. "Molecular Quantum Mechanics", Parts 1 and 2. Clarendon Press, Oxford, 1970.
2. L. Pauling and E. B. Wilson Jr. "Introduction to Quantum Mechanics", Chap. 6, p. 151. McGraw-Hill, New York, 1935.
3. W. Gordy and R. L. Cook. "Microwave Molecular Spectra", Appendix III, p. 647. Wiley, Chichester and New York, 1970.
4. A. Carrington. "Microwave Spectroscopy of Free Radicals", Chap. 3, p. 87. Academic Press, London and New York, 1974.
5. G. Herzberg. "The Spectra and Structures of Simple Free Radicals", pp. 90 and 128. Cornell University Press, Ithaca, 1971.
6. E. B. Wilson Jr., J. C. Decius and P. C. Cross. "Molecular Vibrations", Chap. 2, p. 11. McGraw-Hill, New York, 1955.
7. I. M. Mills. Specialist Periodical Reports: "Theoretical Chemistry", Vol. 1, Chap. 4, p. 118. The Chemical Society, London, 1974.
8. J. Laane. *Quart. Rev.* **25**, 533 (1971).
9. J. D. Lewis and J. Laane. *J. Mol. Spectroscopy* **65**, 147 (1977).
10. J. D. Lewis, T. B. Malloy, T. H. Chao and J. Laane. *J. Mol. Structure*, **12**, 427 (1972).
11. S. I. Chan and D. Stelman. *J. Mol. Spectroscopy*, **10**, 278 (1963).
12. D. R. Herschbach and V. W. Laurie. *J. Chem. Phys.* **40**, 3142 (1964).
13. T. M. Sugden and G. N. Kenney. "Microwave Spectroscopy of Gases", Chaps. 2, 3 and 4. van Nostrand, London, 1965.
14. I. M. Mills. "Molecular Spectroscopy: Modern Research" (K. N. Rao and G. W. Mathews, Eds), Chap. 3, p. 115. Academic Press, London and New York, 1972.
15. L. H. Scharpen. *J. Chem. Phys.* **48**, 3552 (1968).
16. W. Gordy and R. L. Cook. "Microwave Molecular Spectra", Chap. 8, p. 205. Wiley, Chichester and New York, 1970.
17. D. O. Harris, H. W. Harrington, A. C. Luntz and W. D. Gwinn. *J. Chem. Phys.* **44**, 3467 (1966).

3
Experimental Methods of Studying Large Amplitude Internal Motions in Molecules

3.1 Introduction

In Chapter 1 brief reference was made to various experimental approaches which have been used to obtain data relating to internal rotation and other large amplitude vibrations in molecules. If these methods are judged on the quality and quantity of data produced in the literature, then, at the present time, four are outstanding. These are microwave spectroscopy, infrared and Raman spectroscopy, nuclear magnetic resonance spectroscopy and gas phase electron diffraction. In this chapter aspects of these particular techniques are discussed which are directly associated with the study of large amplitude vibrations.

Most students of chemistry or physics encounter these physical methods of studying molecular structure in undergraduate courses and so lengthy descriptions of basic theory and experimental arrangements have not been included here. This information is readily available in many current student textbooks and specialist texts. Instead, a description is given of the relationships between the experimental data obtained by each method and the parameters which describe internal motions in molecules.

3.2 Microwave Spectroscopy

The term microwave spectroscopy usually refers to the study of the pure rotational spectra, and associated spectra, of dipolar molecules in the gas phase at low pressures, using microwave radiation. The microwave region of the electromagnetic spectrum lies between the shorter wavelength radio waves and the far-infrared. Spectroscopic studies on molecules have been made throughout this region although most of the work has been concentrated in the frequency range 8–40 GHz, sources of radiation such as klystrons and backward wave oscillators being commercially available for this part of the spectrum.

Compared to spectroscopy in other regions of the spectrum, microwave spectroscopy is characterized by precise frequency measurement and extremely high resolution. The use of the sources mentioned above together with electronic counting techniques enables typical frequency measurements to be made with an accuracy better than $\pm 0\cdot05$ MHz and, with the line width at half height of a pure rotational absorption line of the order of 1 MHz, lines separated by $0\cdot5$ MHz or less can commonly be resolved. These advantages are, however, offset to some extent by the relative weakness of microwave absorption although this can be largely overcome by the use of special techniques of molecular modulation (e.g. Stark modulation) and electronic amplification.

The basic data obtained from an analysis of measured rotational line frequencies of a molecule are the molecular rotational constants. As shown in Chapter 2, these are inversely proportional to the principal moments of inertia which, in their turn, are related to the molecular structure. In general, non-equivalent rotational isomers, have different principal moments of inertia and hence distinctly different rotational spectra. In instances where rotational isomers are equivalent, the principal moments of inertia, and so the rotational spectrum of each isomer will be identical. Since the rotational constants themselves are a quantum mechanical average over the vibrational state of the molecule different vibrational states of a given rotamer often have well resolved microwave spectra. It is from these spectra of the various vibrational states of separate isomers that information can be obtained about energy differences between the rotamers and between their respective excited vibrational states. Analysis of data of this kind enables potential energy curves governing large amplitude vibrations to be derived.

In practice there are a number of methods by which the relevant information may be extracted from the spectra. The simplest of these in principle is the intensity method.[1] The intensity of a spectroscopic absorption is related to the number of molecules in the lower energy state of the transition concerned and, provided due account is taken of other factors affecting the line intensity such as the dipole moment and statistical weight effects, the relative populations of different molecular states may be obtained from a comparison of the intensities of spectral lines. Assuming that the thermal equilibrium of the system is not destroyed by the radiation, the population ratio between two molecular states is related to the energy separation between these states through the Boltzmann expression (equation (1.2)). Large amplitude vibrations are usually the lowest frequency molecular modes, often $100 \, cm^{-1}$ or less, and at room temperature several excited states may be appreciably populated allowing the energy differences between a number of different vibrational states to be determined.

The energy differences between, for example, the ground vibrational states of different rotamers may be determined in the same way provided of course the rotameric forms are present in sufficient quantity to be detected. The practical limits of detection may vary from case to case, depending on molecular weight, dipole moments and complexity of spectra, but usually the spectrum of a rotamer can be identified if it is present in a concentration greater than about 10%. This figure corresponds to an energy difference between two rotamers of about $5.7 \, kJ \, mol^{-1}$ at room temperature.

An obvious attraction of the intensity method is its basic simplicity. In practice, however, the method suffers rather badly from experimental difficulties in its application, and great care has to be exercised in carrying out intensity measurements on microwave spectra. For example spectrometer response can vary quite dramatically with frequency, and it should be noted that to obtain the population ratios integrated intensities should be compared (see Fig. 3.1). However, it is much more convenient experimentally to compare peak heights and in cases where line shapes are the same, a fact that is

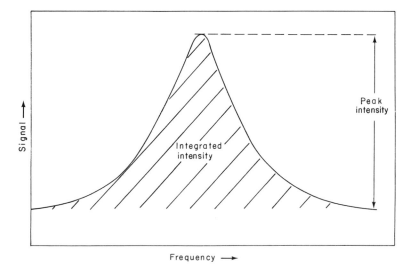

Fig. 3.1   Illustration of the meaning of the terms "Integrated Intensity" and "Peak Intensity" as applied to a spectral line.

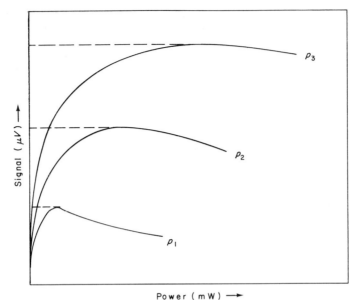

Fig. 3.2   Variation of signal amplitude with incident power level for a typical microwave absorption line for sample pressures $p_3 > p_2 > p_1$.

usually true to a good approximation within the vibrational states of a given rotamer, the ratio of the peak heights compares closely with that of the integrated intensities. A procedure has been developed which allows peak intensity measurements to be used to compare energy level populations even in cases where the line shapes differ. [2] Basically, use is made of the relationship between power saturation of a microwave absorption line and the relaxation processes (e.g. collisions) which tend to oppose saturation. Power levels can be chosen at which peak heights can be shown to be truly representative of molecular population. Typical variations of spectrometer signal with radiation power for several concentrations of an absorbing species have the shapes of the curves shown in Fig. 3.2. The maximum signal amplitude is proportional to the molecular concentration in the lower of the two energy levels involved in the transition. In order to make intensity measurements of this kind special features such as a microwave bridge system, have to be incorporated into the spectrometer. Integrated line intensities can also be measured by digital integration carried out over selected fractions of the line-width with subsequent conversion to the line area on the assumption of a particular line shape, usually Lorentzian. The consequence of the experimetal difficulties in the application of the intensity method is that even with great care, vibrational excitation energies can seldom be determined to an accuracy much better than 5% and experimental errors of 10–20% are more common.

Energy differences between vibrational states and estimates of potential barriers may also be obtained from a knowledge of the way in which rotational constants vary with vibrational state or from the frequency separations between components of rotational absorption lines which are split into multiplets as the result of coupling between internal and overall angular momentum of the molecule. The latter method, often called the frequency splitting method, has been the principal approach used in determining barriers to internal rotation by microwave spectroscopy, and is particularly powerful for internal rotation problems where one of the groups is a symmetric rotor. In the case of a methyl group for example, as shown in Chapter 5, the triply degenerate torsional states are split into a doubly degenerate state (the E state) and a singly degenerate state (the A state) and so each rotational transition of the molecule consists of two components which are often clearly resolved, one associated with the A sub-level and one with the E sub-levels. The $A–E$ splittings are a sensitive function of the barrier height and careful measurement of the frequency separations leads to very accurate barrier heights. The details of these calculations are given in Chapter 5.

Important information relating to vibrational potential functions can often be obtained from an investigation of the variation of rotational constants with vibrational excitation. The rotational constants of a molecule in a particular vibrational state v can be expressed as a power series in the average values of the vibrational coordinate x for that state, viz:

$$B_v = B_e - \alpha_1 \langle x \rangle_v + \beta_1 \langle x^2 \rangle_v + \dots \tag{3.1}$$

where B_e is the rotational constant for the equilibrium configuration and α_1 and β_1 are empirical parameters. In order to use expressions such as equation (3.1) to determine the potential function governing the vibration, various trial potential functions are chosen, the average values of the powers of the coordinates determined, and an attempt made to reproduce the observed variation in the rotational constants with vibrational excitation.

For planar molecules, such as nitrobenzene for example, in which there is a low frequency out of plane vibration, an estimate of the torsional frequency, v_t, can often be obtained from the variation of the inertial defect (Section 2.8.2) with torsional state.[3] This is particularly applicable when the barrier to internal rotation is high. Assuming harmonicity of the vibration the difference between inertial defects in adjacent torsional states is given by,

$$\Delta_{v+1} - \Delta_v = -h/2\pi^2 v_t$$

provided the torsional vibration is much lower in frequency than the other molecular modes. In certain cases frequencies determined by this method can provide a useful check on those determined by intensity measurements.

The commercial development of stable, wide-banded backward wave oscillators as sources of microwave power has enabled rapid scanning of the microwave spectral region to be carried out as a routine measure. For certain molecular structures, viz: near-prolate symmetric tops with Kappa (Section 2.8.3) between -1.0 and -0.7 and possessing a sizeable μ_a dipole component, the low resolution spectra thus obtained have proved to be a valuable means of establishing the presence of more than one molecular conformation, and in certain instances, the approximate shape of the molecular species.[4] Under these low resolution conditions all the μ_a rotational transitions of each $J + 1 \leftarrow J$ spectral jump fall close together under one broad envelope, and the distance apart of these absorption bands is given by the rotational constant $(\overline{B+C})$. This constant differs slightly from $(B + C)$ for the ground vibrational state by an amount which depends on the molecular asymmetry and on vibrational effects. Useful results may be obtained if the value of $(\overline{B+C})$ changes significantly with conformational angle, whereby each rotamer will have its own set of band spectra with a characteristic $(\overline{B+C})$ separation.

Microwave and far-infrared spectra reveal that there are many molecules in which the energy separation between the ground and first excited vibrational states is abnormally small. Ammonia is an example, its microwave spectrum being unusual in that it arises through the "umbrella like" inversion of the molecule (Fig. 3.3) rather than through pure rotation. In this case the first excited state is only 0.8 cm^{-1} above the ground state. This doubling of the vibrational energy levels is characteristic of potential functions with equivalent minima and the separations between the levels are extremely sensitive to the barrier heights and shapes. The ammonia molecule has by no means the smallest vibrational separation known, separations between ground and excited vibrational states of only a few MHz have been measured in several cases. Measurements of this kind are extremely important pieces of data in the characterization of large amplitude vibrations and their associated potential functions and it is unfortunate that circumstances are often such that only the separations between the very lowest vibrational states can be obtained by microwave spectroscopy itself leading to a rather poorly defined potential function. The situation is usually improved by the inclusion of data pertaining to higher vibrational states obtained by other techniques.

Microwave spectroscopy is of course an extremely powerful method of

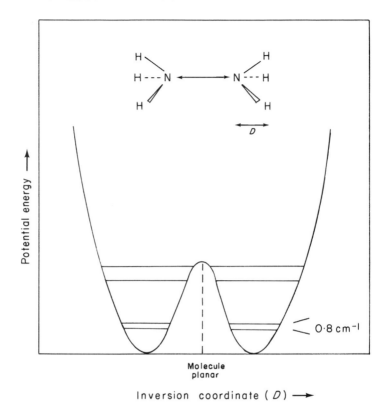

Fig. 3.3   Inversion in ammonia: illustration of the shape of the potential function governing the vibration and the associated inversion levels.

determining accurate molecular structures, and in a number of cases, 3-fluoropropene is an example,[5] it has been shown that small but significant structural differences may exist between non-equivalent rotamers. Thus, in some cases, it may not be possible to interconvert non-equivalent rotamers simply by rotation around a chemical bond and this must be reflected in the potential function governing the vibration.

While it is clear that microwave spectroscopy is, by itself, a very sensitive and informative experimental probe of these vibrational problems, the data obtained is often best used in conjunction with data from other sources, particularly far-infrared spectroscopy. When used in this context microwave spectroscopy has contributed much of the most reliable and accurate data available at the present time on large amplitude molecular vibrations and their associated potential functions.

3.3 Infrared and Raman Spectroscopy

Infrared and Raman spectroscopy are included in the same section because they provide molecular data of a very similar nature. Having stated this it must be pointed out that the techniques themselves are rather different, infrared spectroscopy usually being straightforward absorption spectroscopy, whereas Raman spectroscopy involves an analysis of radiation scattered inelastically from a sample. One of the main results of the interaction of the incident radiation with the sample in both cases is a change in vibrational energy of the molecules in the sample. In the case of infrared spectroscopy only those vibrations which involve a change in the molecular dipolar moment are active, whereas for Raman activity a change in molecular polarizability during the vibration is essential. The techniques are therefore to a large extent complementary and when used conjointly represent a very powerful method of studying the movements of atoms in molecules.

The infrared spectral region is usually defined in wavenumber units as 10^4–10 cm^{-1} and commerical spectrometers generally cover the region from 5000 cm^{-1} down to 200 cm^{-1} called the mid-infrared. The region below 200 cm^{-1}, the far-infrared, is traditionally much more difficult to cover experimentally mainly because of the lack of strong radiation sources and the intense atmospheric absorption which occurs at these wavelengths. Modern developments in the form of interferometric techniques and better detectors have greatly improved the coverage of this region and spectra down to 10 cm^{-1} or less can nowadays be obtained readily. In Raman spectroscopy the main experimental difficulties were, until relatively recently, associated with the weakness of the spectra. With the advent of laser sources for Raman spectrometers this problem has been largely overcome and the applicability and usefulness of the technique significantly enhanced. In Raman spectroscopy coverage of the region from 4000–50 cm^{-1} can generally be achieved without difficulty. Both infrared and Raman spectra can be obtained from vapour, liquid or solid samples, the quantity of sample required being governed by factors such as cell size and the nature of the sample itself.

Measurement of frequencies of the vibrational transitions of a molecule can usually be made with an accuracy of $\pm 1 \text{ cm}^{-1}$ or less and provides a direct determination of the separation between the vibrational energy levels involved in the transitions. The vibrational frequencies themselves are characteristic of the spatial arrangements of the nuclei, the atomic masses and the forces acting between the nuclei which tend to restore the molecule to its equilibrium configuration during vibrational distortion.

Consequently several of the vibrational transition frequencies of different rotamers of a particular molecule may well be different and, since a resolution of about 0.2 cm^{-1} is commonly attainable in infrared spectroscopy, the presence of more than one rotamer may be revealed by a doubling of some of the absorption bands. In the liquid phase the relative intensities of the bands due to separate rotamers often alter in solvents of different polarities provided the rotamers have significantly different dipole moments. In this way specific absorption bands may be assigned to the more polar or less polar rotamer.

Vapour phase absorption bands are usually fairly broad and often have characteristic shapes arising from unresolved fine structure. The fine structure is due to changes in rotational energy accompanying a vibrational change and useful structural information can often be derived from the shape of the band envelope. More precise structural data can, of course, be obtained when the rotational fine structure can be adequately resolved.

In the condensed phases the rotational freedom of molecules is restricted and vibrational bands are almost invariably narrower than in vapour phase spectra. It is sometimes found, however, that in molecules containing a methyl group for example, which has a low barrier to internal rotation some of the internal rotational motion is retained in solution. This is revealed in the infrared spectrum through the presence of unusually broad absorptions for vibrations having an associated change in dipole moment perpendicular to the axis of the methyl group.[6] There is also evidence that this internal rotation may be retained to some degree even in the solid phase and a study of the effect of temperature on the widths of these bands can provide an estimate of the height of the barrier hindering the internal rotation under these circumstances.[7]

Strong evidence for the existence of more than one rotational isomer of a molecule can often be obtained by cooling the vapour or liquid sample in a low temperature cryostat. A polycrystalline sample can be obtained in this way and, in the infrared spectrum of such a sample, it is often found that some absorption bands, prominent in the vapour or liquid phase spectra, are completely missing, Fig. 3.4. Under the low temperature conditions the thermodynamically most stable form of the molecule predominates and the spectra from less stable conformers disappear. Matrix isolation techniques,[8] where the sample is trapped in an inert, transparent matrix such as an inert gas, are often used in this context to remove the effects of crystal and dipolar interactions which sometimes complicate the spectrum of the solid.

Energy differences between rotamers can be conveniently determined by studying the effect of temperature on the relative intensities of the absorption

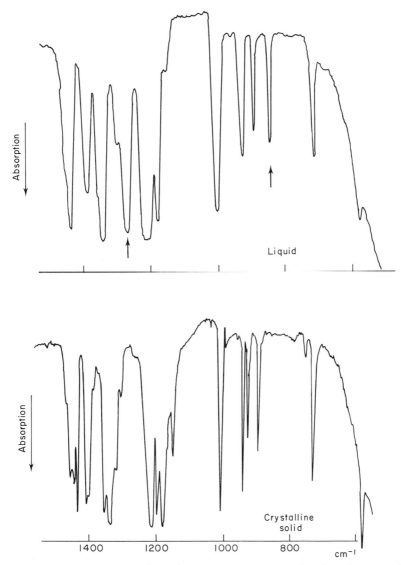

Fig. 3.4   Infrared spectra of methyl cyanoacetate as (a) liquid at ambient temperatures, and (b) crystalline solid at 77 K. The absorptions marked in spectrum (a) are missing in (b) and are associated with the thermodynamically less stable rotamer of the molecule.

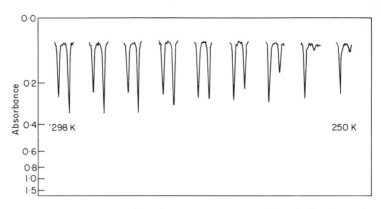

Fig. 3.5   The effect of varying the temperature on the relative intensity of the 893/845 cm $^{-1}$ rotameric doublet in the infrared spectrum of methyl cyano-acetate.

bands due to each rotamer, Fig. 3.5. Use is made of Beer's Law which relates the concentration of an absorbing species (c) to its optical density ($\log_{10} I_0/I$),

$$\log_{10} \frac{I_0}{I} = \alpha l c \qquad (3.2)$$

where I_0 and I are the intensities of incident and transmitted radiation respectively corresponding to a particular absorption, l is the path length and α is the extinction coefficient. This may be combined with the Boltzmann Law (equation (1.2)) to give,

$$\ln \left(\frac{\text{optical density 1}}{\text{optical density 2}} \right) = -\frac{\Delta E}{RT} + \ln g + \ln \frac{\alpha_1}{\alpha_2} \qquad (3.3)$$

where g is the ratio of the statistical weights of the two rotational isomers 1 and 2. It is usual to assume that the third term on the right-hand side of equation (3.3) is to a large extent temperature independent and hence a graph of the left-hand side of equation (3.3) against T^{-1} results in a linear plot from which ΔE, the required energy difference, is readily obtained. To a good approximation ΔE can be equated to ΔH for these rotameric systems and, in favourable cases, the accuracy with which ΔH may be determined is better than 10%. Although integrated band intensities should be used in these measurements the method gives reliable results using peak intensities provided the lines are reasonably well resolved. The same procedure can be used to determined ΔH values from Raman spectra, however, in this case, since scattered radiation is involved, the band intensities are directly proportional to the concentrations of the scattering species.

The fundamental absorption of many large amplitude vibrations occurs in the far-infrared region of the spectrum, i.e. below $200\ \mathrm{cm}^{-1}$, and the identification and measurement of absorption bands associated with these vibrations can lead to estimates of barriers hindering internal motions. Often, data obtained in this way can be directly compared with vibrational energy level separations estimated by other methods (e.g. microwave spectroscopy). One of the major difficulties encountered in the application of this approach to the determination of barrier heights is that the correct assignment of weak bands in infrared spectra is notoriously difficult, and in many cases large amplitude vibrations give rise to only very weak absorptions. Methyl group torsions, for example, when active at all, are often exceedingly weak ($d\mu/d\alpha$ for these vibrations is small) and their assignment consequently dubious. In this respect Raman spectroscopy can sometimes be more definitive. Polarization measurements can be made on lines and used to add weight to the assignment of a weak band. In the solid phase torsional bands are often found to be somewhat stronger than in the vapour or liquid phase spectra.[9] However, they are generally shifted to higher frequency making it possible only to place an upper limit on the height of the barrier hindering the torsion.

Sometimes "hot" bands of a torsional vibration can be observed involving vibrational transitions between excited vibrational states, e.g. $v = 2 \leftarrow 1$, $3 \leftarrow 2$ transitions. This information about higher vibrational states is obviously of great value in constructing a meaningful potential function governing the motion. An alternative, and sometimes convenient method of measuring torsional vibrations is to observe them in combination with other absorptions in the near-infrared; ring puckering vibrations have also been identified in this way.

3.4 Nuclear Magnetic Resonance Spectroscopy

Nuclear magnetic resonance spectroscopy is usually concerned with the study of the absorption of radio frequency radiation by a sample placed in a strong, homogeneous magnetic field. When the correct radiation frequency is applied for a given magnetic field strength, transitions occur between the energy levels associated with the allowed orientations of the magnetic moments of the nuclei in the sample relative to the direction of the magnetic field. The absorptions which occur are inherently very weak and considerable electronic amplification is needed for their detection. The exact value of the magnetic field experienced by a given nucleus in a molecule is slightly dependent on its molecular environment and so the resonance conditions for otherwise identical nuclei in different parts of a molecule may vary. Although these differences are small, of the order of one part in 10^5 for protons for example, it is this effect which makes nuclear magnetic resonance spectroscopy particularly useful in the study of molecular structure.

The line-width at half-height of a magnetic resonance absorption can be less than 1 Hz and it is often desirable to resolve absorptions separated by this small amount. To achieve this resolution, extremely stable and homogeneous magnetic fields are required and the sample itself is usually spun to reduce the line broadening which arises from any small inhomogeneities which exist in the field. In a normal magnetic resonance experiment, provided saturation of the signal is avoided, the area under an absorption line is proportional to the number of resonating nuclei present and inversely proportional to the temperature. Usually electronic integration is used to provide relative intensities of the absorption lines in the spectrum with an accuracy of the order of 5 %. The method, however, becomes unreliable when overlapping peaks occur in the spectrum. In cases where the line shapes are known to be identical, straightforward comparison of peak heights gives a measure of intensity.

For molecules in which nuclei are changing their environment as a result of some internal motion the signals corresponding to different stable nuclear configurations may be separated by only a few hertz. If it is assumed that the different configurations of the nuclei have the same lifetime δt, and provided the signals from these configurations do not overlap, the uncertainty principle predicts that the lines will be broadened by,

$$\delta v \simeq (2\pi\delta t)^{-1}$$

Consequently the shapes and widths of the absorption lines in a magnetic resonance spectrum might be expected to be affected in a noticeable way when the lifetimes of the nuclei in their separate environments are in the range $\sim 10^{-1}$–10^{-5} s. Thus information about processes such as internal rotation, inversion, ring inversion motions, etc., all of which may lead to changes in the molecular environment of certain nuclei, is potentially available from a study of the shapes of the absorption lines associated with the moving nuclei. The line-shape theory appropriate to this problem is complex and will not be given here.

In molecules where the potential barrier hindering the movement of the nuclei is high enough to effectively isolate them in their separate environments at a given temperature, and, provided the nuclear population in a given environment is sufficiently high to be detected ($\sim 5\%$), the spectrum at that temperature consists of a superposition of the separate signals from each nuclear environment. Where different rotamers are present the energy difference between the two forms may be estimated from the variation of the relative intensitity with temperature of the spectrum of each form. In the other extreme where the barrier hindering the nuclear motion is less than about 20 kJ mol^{-1}, exchange between the separate nuclear environments may occur very rapidly compared to the critical rate. Under these circumstances the absorption line associated with these nuclei becomes a weighted average of the signals from each separate environment.

From the point of view of nuclear magnetic resonance spectroscopy the most interesting cases lie between these two extremes when the barrier to the motion is in the range ~ 20–150 kJ mol^{-1}. Under these conditions the line shape may vary dramatically with temperature as the exchange rate between two molecular sites alters. Figure 3.6 shows how the line shape may alter as a function of temperature. At low temperatures when the exchange

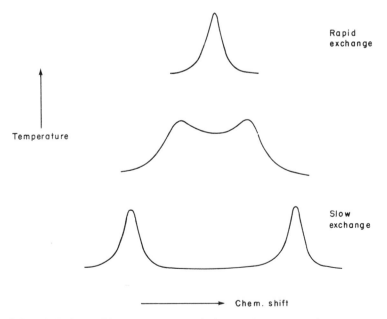

Fig. 3.6   Variation with temperature of the nuclear magnetic resonance line shape for nuclei exchanging between two equally populated sites.

rate is assumed to be small, separate signals from each site are obtained although the lines may be broadened somewhat from the limit of zero exchange. As mentioned previously, this broadening can be related through the uncertainly principle to the probability per unit time ($\tau \equiv \delta t^{-1}$), that a nucleus in either site will exchange to the other molecular site. As the temperature is raised so the lines continue to broaden and converge until eventually they coalesce in the limit of rapid exchange. The values of τ for a given process may be obtained from linewidth measurements in the limit of slow or rapid exchange. However, in the transition period, the data must be extracted from the spectrum by comparing calculated line shapes for various τ values with experimental line shapes. In this way values for a series of temperatures may be obtained and exchange can then be treated in a similar manner to a simple rate process with rate constant k proportional to τ, and the barrier hindering the exchange obtained from the Arrhenius equation.

This approach has been used extensively to study molecules with barriers in the appropriate range. However, it often suffers from the fact that for these molecules the collapse of the chemical shifts occurs over small temperature ranges and consequently the derived Arrhenius parameters are somewhat inaccurate. One way of overcoming this difficulty is to use the spin-echo method.[10] In applying the spin-echo technique the r.f. field is pulsed at high power and at short duration compared with the relaxation times of the nuclei. It is found that the decay of the signal received at the detector in such an experiment can be related to the relaxation times and, in cases where internal motion is occurring, to τ values. Thus a study of the signal decay leads to τ values which may be used in an analysis similar to that described above to give the height of the barrier hindering the internal motion. The technique is applicable over a larger range of temperature and rates than the line-shape method and, with care, can provide an improvement in accuracy in some cases.

Barriers hindering internal rotation in molecules in the solid phase can also be obtained by nuclear magnetic resonance techniques.[11] The spin-lattice relaxation times of the nuclei, usually measured by pulse techniques, may be related in these cases to the correlation time, which is the time taken for the internal top to rotate through one radian. If the correlation time (τ_c) can be determined at various temperatures then an equation of the form,

$$\tau_c = \tau_0 \, e^{-E_A/RT} \tag{3.4}$$

where τ_0 is constant, can be used to calculate the height of the barrier hindering internal rotation. Barriers from a few kJ upwards have been measured by this method.

Recently, information about internal rotation has been obtained from studies of the spectra of molecules partially orientated in a liquid crystalline solvent.[12] The effect of the high degree of orientational order present in the nematic liquid crystalline phase results in partial orientation of the solute molecules. Consequently the resulting NMR spectrum is affected strongly by the magnetic dipole–dipole interactions which are normally averaged to zero for non-viscous liquids and which completely dominate the corresponding spectra of the solid phase. In the liquid crystalline state, however, molecular diffusion occurs fairly readily and this ensures that only *intra* molecular dipole interactions are non-zero. The resulting NMR spectral pattern for solutes is interpreted in a manner similar to conventional NMR spectra except for the additional dipolar interaction term. This term contains parameters which are functions of molecular geometry, suitably averaged over vibrational motions. Trial orientation parameters and nuclear coordinates are used as the starting point of a least squares fitting procedure which converges on to the experimental values for the coupling constants in the molecule thus enabling the shape of the nuclear skeleton to be determined. The probability distribution of the nuclei comprising the internal top (e.g. a methyl group) is obtained using a simple cosine potential function and assuming a Boltzmann distribution in the vibrational energy states. This probability distribution is included in the least squares procedure and the value of V_3 chosen which gives the best fit to the experimental parameters. The barrier heights obtained are, at present, rather inaccurate when compared with those derived from microwave spectroscopy for example, however, the method does allow internal rotation to be studied in molecules with barrier heights of only a few kJ mol^{-1}.

3.5 Gas Phase Electron Diffraction

In the gas phase electron diffraction experiment a beam of electrons with an energy of several kV is directed at a jet of sample vapour. The electrons are scattered by the sample and the scattered beam allowed to fall on to a photographic plate. The result is a series of concentric diffraction rings on the plate and it is from this pattern that molecular data is derived. The diffraction pattern arises from the scattering of the electron beam by different parts of the individual molecules in the gas sample and, since there is a large number of molecules oriented randomly with respect to the incident beam, it represents an averaged pattern. The intensity $I_t(s)$ of the scattered beam is a function of the scattering angle (where s is a parameter related to the scattering angle) and may be written as the sum of two separate parts,

$$I_t(s) = I_a(s) + I_m(s) \tag{3.5}$$

$I_a(s)$ is the intensity derived from atomic and incoherently scattered electrons and contains no molecular information. $I_m(s)$ is the molecular scattering intensity and it is this component of the scattered radiation which is important for the determination of molecular parameters. In order to concentrate on the $I_m(s)$ term the $I_a(s)$ term is removed from the expression for the total scattering intensity by the subtraction of an analytical expression for $I_a(s)$ or, more usually, an empirical atomic scattering curve which also takes account of effects such as scattering from parts of the apparatus and multiple scattering within the molecule. This empirical curve is sometimes referred to as the background curve.

The important molecular parameters which occur in the analytical expression for $I_m(s)$ are the interatomic distances in the molecule, both bonded and non-bonded, and the mean square amplitudes of vibration of the atoms. In cases where the molecular vibrations can be assumed to be approximately harmonic and no very large differences occur between the atomic numbers of the atoms in the molecule, these molecular parameters can be obtained by using a least squares iterative procedure to fit the observed and calculated molecular scattering curves. In cases where these approximations are invalid, a modified expression for $I_m(s)$ has to be used and the procedure for the extraction of interatomic distances and mean amplitudes of vibration is more involved.

Instead of working only in terms of the molecular scattering curve $I_m(s)$ itself, it is often convenient to make use of the corresponding radial distribution curve. This curve is the Fourier transform of a slightly modified version of $I_m(s)$ and when the radial distribution curve is plotted as a function of distance it is found to have maxima at distances corresponding to interatomic distances in the molecule. The intensity of a peak in the radial distribution curve is a function of the atomic numbers of the atoms involved, the interatomic distance, and the number of times that distance occurs in the molecule. The peak width at half height is related to the mean amplitudes of vibration of the atoms. Problems often occur in the analysis of radial distribution curves owing to overlapping peaks, and it is usual to resolve these by fitting the observed peak shape to the sum of Gaussian components using spectroscopically determined values of the mean amplitudes of vibration when necessary.

Application of electron diffraction to the study of internal motions[13] is most straightforward when stable molecular conformers separated by high barriers exist. Under these conditions the radial distribution curve for the molecule consists simply of a weighted sum of the radial distribution curves for the individual conformers and so, by trial and error procedures or by using a least squares fitting technique on the observed and calculated molecular scattering curves, or sometimes the radial distribution curves, the proportions of each conformer present can be derived. The energy differences between the conformers can be estimated in a straightforward way, usually with an accuracy of, at best 5–10%, by comparing the relative integrated intensities of the outer peaks in the radial distribution curves arising from each conformer and using an equation of the form of (1.1) to obtain the required energy difference.

In cases where the barrier height falls below about 40 kJ mol^{-1}, it becomes no longer valid to consider a torsional vibration for example as a small amplitude vibration and the analysis of the diffraction pattern becomes more complex. The usual approach, in say the case of internal rotation, is to synthesize the molecular intensity curve or the radial distribution curve by taking a weighted sum of a very large number of rotamers as the internally rotating group is rotated through 360°. As in the case case of most analyses of large amplitude vibrations it is generally assumed that the large amplitude mode can be isolated from all other molecular vibrations, the other vibrations being taken to be harmonic.

As shown in Fig. 3.7 internal rotation by an angle α alters the internuclear distance R_{ij} in a molecule from its equilibrium value R_{ij}^e. The essence of the method is to determine a probability distribution for the distance R_{ij} and then to use this to calculate the contribution of the atom pair ij to the molecular scattering intensity. The required probability distribution, and hence the contribution by atom pairs ij to the overall molecular scattering involves a weighting factor which is directly governed by the potential function hindering the internal rotation. The form of this weighting factor is generally derived by inserting the chosen potential function, usually a simple cosine potential, into the wave equation for the torsion, and determining the eigenvalues and eigenfunctions of the torsional vibration. The eigenfunctions can then be used with the assumption of a Boltzmann distribution of molecules in the torsional states, to give the necessary weighting factor. By making use of mean square amplitudes of vibration from spectroscopic or other sources, the contribution from distances between atom pairs i and j to the molecular scattering intensity can then be calculated and the Fourier coefficients, V_n, varied to give a good fit to the observed scattering intensity and radial distribution curves.

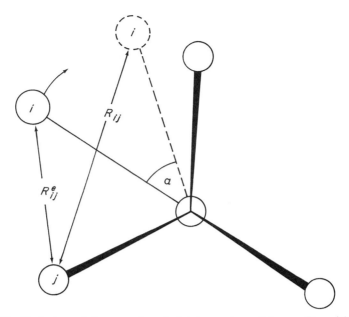

Fig. 3.7   Variation of the non-bonded internuclear distance R_{ij} with internal rotation α seen in projection for a threefold symmetric top rotating against an asymmetric frame (e.g. CH_3OCl) viewed along the O—C bond).

Potential functions derived in this way often suffer from rather high uncertainty limits, barrier heights having error limits of 30% in many cases. However, the technique is applicable to many cases which are inaccessible or very difficult to study spectroscopically owing for example to lack of a dipole moment or inactivity in infrared or Raman spectra, and if the accuracy of the results can be significantly improved, electron diffraction will undoubtedly provide a valuable complement to the spectroscopic methods in the study of large amplitude vibrations.

Further Reading

1. J. C. D. Brand, J. C. Speakman and J. K. Tyler. "Molecular Structure". Arnold, London, 1976.
   This is a modern account of the main spectroscopic and diffraction methods of studying molecular structure.
2. N. B. Colthup, L. H. Daly and S. E. Wiberley. "Introduction to Infrared and Raman Spectroscopy". Academic Press, London and New York, 1964.
   An excellent introduction to both practical and theoretical aspects of vibrational spectroscopy.

3. W. Gordy and R. L. Cook. "Microwave Molecular Spectra". Wiley, Chichester and New York, 1970.
   A comprehensive text on pure rotational spectroscopy.
4. J. W. Emsley, J. Feeney and L. H. Sutcliffe. "High Resolution Nuclear Magnetic Spectroscopy", Vols. I and II. Pergamon Press, Oxford, 1965.
   A two volume treatise including both theoretical and practical aspects of NMR.

References

1. A. S. Esbitt and E. B. Wilson Jr. *Rev. Sci. Instr.* **34**, 901 (1963).
2. H. W. Harrington. *J. Chem. Phys.* **46**, 3698 (1967); **49**, 3023 (1968).
3. D. R. Herschbach and V. W. Laurie. *J. Chem. Phys.* **40**, 3142 (1964).
4. (a) W. E. Steinmetz. *J. Amer. Chem. Soc.* **96**, 685 (1974).
   (b) M. S. Farag and R. K. Bohn. *J. Chem. Phys.* **62**, 3946 (1975).
5. P. Meakin, D. O. Harris and E. Hirota. *J. Chem. Phys.* **51**, 3775 (1969)
6. W. J. Jones and N. Sheppard. *Proc. Chem. Soc.* 420 (1961).
7. A. V. Rakov. *Opt. Spectry.* **13**, 203 (1962).
8. H. E. Hallam (Ed.). "Vibrational Spectroscopy of Trapped Species". Wiley, Chichester and New York, 1973.
9. J. R. Durig, J. Bragin, S. M. Craven, C. M. Player Jr. and Y. S. Li. Torsional frequencies and barriers to internal rotation from far infrared spectra of solids, *Develop. App. Spectroscopy*, **9**, 23 (1971).
10. (a) J. W. Emsley, J. Feeney and L. H. Sutcliffe. "High Resolution Nuclear Magnetic Resonance Spectroscopy", Vol. 1, Chaps. 3, 4 and 9. Pergamon Press, Oxford, 1965.
    (b) J. A. Pople, W. G. Schneider and H. J. Bernstein. "High Resolution Nuclear Magnetic Resonance", Chap. 2. McGraw-Hill, New York, 1959.
11. A. M. I. Ahmed, R. G. Eades, T. A. Jones and J. P. Llewellyn. *J. Chem. Soc. Faraday Trans. II*, **68**, 1316 (1972).
12. J. W. Emsley and J. C. Linden. "NMR Spectroscopy using Liquid Crystal Solvents". Pergamon Press, Oxford, 1975.
13. A. H. Clark. Electron diffraction studies and rotational isomerism, Chap. 10 *in* "Internal Rotation in Molecules" (W. J. Orville-Thomas, Ed.). Wiley, Chichester and New York, 1974.

4
The Origin of Potential Barriers

4.1 Introduction

Almost from the time of the discovery of hindered rotation in ethane the origin of rotational barriers has presented a challenging problem to theoreticians, and over the years a variety of explanations has been put forward. An idea of the difficulties involved may be obtained from Table 4.1 where it can be seen that the rotational barrier in ethane is an extremely small fraction

Table 4.1 Comparison of the total molecular energy for ethane with the carbon–carbon bond energy and the barrier height to internal rotation

	kJ mol^{-1}	% of (a)	% of (b)
(a) Total molecular energy of ethane	$2\cdot1 \times 10^5$	—	—
(b) Carbon–carbon bond energy	$3\cdot7 \times 10^2$	$1\cdot8 \times 10^{-1}$	—
(c) Barrier to internal rotation	$1\cdot2 \times 10^1$	$5\cdot7 \times 10^{-3}$	$3\cdot2$

of the total molecular energy and is of the order of three per cent of the dissociation energy of the carbon–carbon bond. In spite of this, considerable progress has been made in recent years in computing barriers for relatively small molecules using molecular orbital techniques and this topic will be discussed later in this chapter. Earlier work has tended to be rather empirical and qualitative in nature and has often sought to explain barriers and conformational energy differences in terms of special features of the bond about which the internal rotation is taking place or in terms of specific forces between non-bonded atoms and groups.[1] This approach may be divided into attempts to account for what may be described as normal barriers and to account for exceptions to expected trends in barriers. Both will be briefly considered in turn.

4.2 Ethane and Early Theories of the Origin of Barriers

The barrier of approximately $12{\cdot}0 \text{ kJ mol}^{-1}$ in ethane may be regarded as the normal value for the threefold barriers preventing the internal rotation of a methyl group when it is bonded to another sp^3 hybridized carbon atom. In this chapter some of the early attempts to account for such barriers will be discussed, because in spite of the success of the *ab initio* computations described in Section 4.6.2, these older methods are still frequently used to rationalize experimental barrier heights and conformational energy differences. The first quantum mechanical approach to the origin of barriers used the valence bond theory which, in its simplest form, leads to the conclusion that the rotational barrier in ethane should be extremely low. Higher approximations were considered and the first of these explored the possibility that the carbon–carbon bond did not have cylindrical symmetry. A serious argument against this approach is that it leads to potential minima at the eclipsed conformations. The effect of modifying the orbitals on the carbon atoms which form the carbon–hydrogen bonds has also been tried by incorporating some 3d character, whereby the three carbon–hydrogen bond orbitals are found to no longer have an electron density which is cylindrically symmetrical. The barrier to internal rotation could then arise from the interactions of such distributions on the two carbon atoms. This approach has been advocated strongly by Pauling, however recent developments indicate that it is not really tenable, although it has an intuitive appeal to many chemists.

4.3 Barriers and Conformational Energy Differences Dominated by Specific Interactions

4.3.1 Different types of intramolecular forces

If the rotational isomerism in a certain molecule differs significantly from that in related molecules it is usual to seek an explanation in terms of intramolecular forces which are present in the molecule being considered but absent in the others. In doing this it is assumed that the conformational energies and barriers can be expressed as sums of terms due to different types of forces.[2] The forces giving rise to the threefold barriers mentioned in the previous section may be included under the general term "barrier forces" and in a series of related molecules these will give very similar contributions. Other types of forces which may be invoked in special situations include:

1. double bond character due to resonance
2. hydrogen bonding
3. steric repulsion

and some examples will now be considered in order to illustrate the importance of these forces in various circumstances.

4.3.2 Double bond character due to resonance

This is particularly important in situations where a single bond occurs between two double bonds, such as in butadiene and related molecules. In a valence bond picture the central bond in these molecules acquires some double bond character because of contributions from structures such as,

to the overall state of the molecule. Molecular orbital treatments lead to similar conclusions about the central bond acquiring double bond character. This double bond character is sufficient to lead to *cis–trans* isomerism in a number of molecules and to potential functions which are dominated by twofold rather than threefold terms. The rotational barriers are found to be somewhat higher than that in ethane but less than that in ethylene.

Conjugation effects between electron rich groups (e.g. OCH_3, NH_2) and unsaturated bonds can also lead to single bonds having partial double bond character. For example in carboxylic esters, the central carbon–oxygen bond has quite different properties to the oxygen–alkyl bond, and this is attributed to interaction of the lone pair electrons on the oxygen atom with the π electron system of the carbonyl group. The barrier to internal rotation about the central carbon–oxygen bond V(a) in most simple esters is between 25 and 50 kJ mol^{-1}, whereas the barrier about the O—R′ bond V(b) is about 4 kJ mol^{-1} and largely threefold.[3]

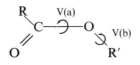

In amines and amides the double bond character in the C—NH$_2$ bond raises the barrier to internal rotation, but the consequent tendency towards a planar arrangement of atoms around the nitrogen lowers the barrier to inversion of the NH$_2$ group. In aniline, contributions from the resonance structures I–III lead to a barrier to inversion of 5·4 kJ mol^{-1} compared to 25·3 kJ mol^{-1} in ammonia.

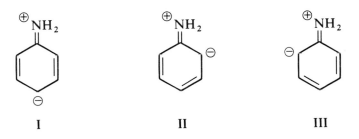

I   II   III

In formamide the contribution from the structure

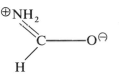

is large enough to lead to a planar molecule.[4] This has important consequences in the geometry of the polypeptide link and the structure of proteins (Chapter 9).

4.3.3 Hydrogen bonding

Intramolecular hydrogen bonding has been invoked to explain conformational behaviour and energies in molecules ranging from simple acids and alcohols to large and complicated systems such as proteins, nucleic acids and carbohydrates. This type of interaction usually arises when hydrogen is bonded to an electronegative atom such as nitrogen or oxygen, and is close to a centre of negative charge such as an atom with lone pairs of electrons, a multiple bond or a conjugated ring system. The nature of the hydrogen bond has been the subject of much discussion and it is now generally recognized to consist of four different components, i.e. electrostatic, delocalization, dispersion and repulsive interactions, and it is believed that these make contributions of similar magnitude to the energy of the bond.

In simple derivatives of methanol of the type CX$_2$YOH which, by analogy

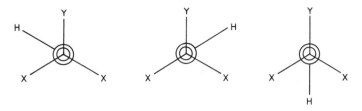

Fig. 4.1   Possible rotamers for a CX_2YOH molecule.

with the parent molecule, may be expected to have three conformations (Fig. 4.1) of almost equal energy separated by barriers of approximately 4.5 kJ mol^{-1}, the effects of hydrogen bonding may be seen from Table 4.2. Ethanol and isopropanol, in which intramolecular hydrogen bonding is unlikely, show the expected rotational isomerism and barriers to internal

Table 4.2   Stable rotamers and barriers to internal rotation about the C—O or C—S bonds for some CX_2YOH and CX_2YSH molecules

	Stable rotamers	Δ^a (cm^{-1})	Barriers to internal rotation (kJ mol^{-1})
Methanol	staggered	—	$V_3 = 4.48$
Ethanol	{ trans, gauche	5.23	$V_3 = 5.0$
Ethyl mercaptan	{ trans, gauche	—	—
Isopropanol	{ trans, gauche	1.53	$V_3 = 7.0$
Isopropyl mercaptan	{ trans, gauche	0.02	$V_3 = 8.3$
Propargyl alcohol	gauche	21.49	{ cis 1.0, tans 8.0
Propargyl mercaptan	gauche	0.230	{ cis 5.6, trans 23.9
Hydroxyacetonitrile	gauche	3.69	{ cis 3.7, trans 4.6
Cyclopropanol	gauche	0.137	{ cis 9.2, trans 18.9

[a] Δ is the tunnelling splitting of the lowest torsional state of the *gauche* rotamer.

rotation. In the remaining molecules only the two equivalent *gauche* rotamers have been positively identified and a relatively low *cis* barrier and a high barrier at the *trans* position have been found. Such behaviour may be readily understood if there is competition between a weak hydrogen bond which favours a *cis* conformation and the factors which lead to the staggered conformations shown in Fig. 4.1 for ethanol and isopropanol. Hydrogen bonding is also important in determining the equilibrium conformations of derivatives of ethanol which are substituted in the methyl group. In chloroethanol, for example, of the nine possible conformations associated with internal rotation about the carbon–oxygen and carbon–carbon bonds only one pair of equivalent rotamers, in which the hydroxyl hydrogen is closest to the chlorine atom, has been observed. Another interesting example of intramolecular hydrogen bonding is found in methoxyacetic acid (Fig. 4.2(a)), where the O—H group of the carboxyl group is orientated at 180° to its normal position as found in formic and acetic acids (Fig. 4.2(b)) and thus forms a hydrogen bond to the ether oxygen atom resulting in a planar five membered ring structure.[5]

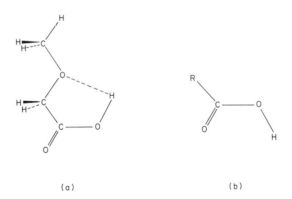

(a)   (b)

Fig. 4.2   (a) The most stable rotamer of methoxyacetic acid, showing the effect of hydrogen bonding. (b) The normal conformation of carboxylic acids.

4.3.4 Steric effects

Steric repulsions occur when two atoms are separated by a distance less than the sum of their van der Waal's radii, and may be expected to play an important part in determining rotational isomerism and barriers in instances where bulky atoms or groups are attached to the atoms of the bond about which internal rotation is taking place. A simple example where such an effect causes significant deviations from a symmetrical threefold potential is ethyl formate.

Methyl formate has the equilibrium conformation shown in Fig. 4.3(a) and a rotational barrier of about 5 kJ mol^{-1} about the O—CH$_3$ bond. Ethyl formate exists as the *trans* and *gauche* rotamers shown in Fig. 4.3(b) and (c) with the *trans* rotamer being more stable by 0·8 kJ mol^{-1}.[6] One consequence of the steric interaction is to lead to the *gauche* rotamers having a dihedral angle of 95° as measured from the *trans* rotamer, instead of the expected 120°. The barrier separating the *gauche* and *trans* rotamers of 4·6 kJ mol^{-1} is rather similar to that in methyl formate but the barrier of about 22 kJ mol^{-1} at the *cis* position is considerably higher.

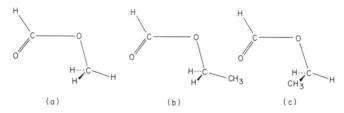

(a)   (b)   (c)

Fig. 4.3   The most stable rotamers of (a) methyl formate, (b) and (c) ethyl formate.

A more dramatic example of the effect of steric hindrance on internal rotation barriers is found when methyl formate is compared with methyl vinyl ether (Fig. 4.4). The barrier to internal rotation of the CH$_3$ group in methyl vinyl ether is about 16 kJ mol^{-1} compared to the value of 5 kJ mol^{-1} in methyl formate, clearly illustrating the importance of the steric contribution to the barrier.

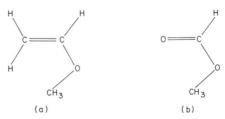

(a)   (b)

Fig. 4.4   The most stable conformations of (a) methyl vinyl ether (b) methyl formate.

4.4 The Derivation of Information about Intramolecular Forces from Experimental Potential Functions

Experimental potential functions may be used in a number of ways to give information about intramolecular forces and in this section one of these methods will be illustrated by the simple example of fluoroacetyl fluoride.[7]

In the gas phase this molecule exists as two rotamers, the more stable *trans* form and the less stable *cis* form shown in Fig. 4.5. The energy difference between the two rotamers is $3\cdot8$ kJ mol^{-1} and there is a high barrier of approximately 20 kJ mol^{-1} separating them, the maximum occurring at an angle of about 80° from the *trans* position. The related molecules acetyl fluoride and trifluoroacetyl fluoride have threefold barriers of 4.4 and $5\cdot8$ kJ mol^{-1} respectively.

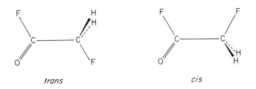

trans   *cis*

Fig. 4.5   The stable rotamers of fluoroacetyl fluoride.

The interaction of an atom A in the CH_2F group of fluoroacetyl fluoride with an atom B in the COF group may be expressed as a Fourier series in the torsional angle α.

$$V^{AB}(\alpha) = V_1^{AB} \cos\alpha + V_2^{AB} \cos 2\alpha + \ldots$$

In particular the interaction of the F atom with the COF group is,

$$V^{F\cdots COF} = V^{FO}(\alpha) + V^{FF}(\alpha + \pi)$$
$$= (V_1^{FO} - V_1^{FF}) \cos\alpha + (V_2^{FO} + V_2^{HF}) \cos 2\alpha + \ldots$$

Similarly the interaction between one of the hydrogen atoms and the COF group is,

$$V^{H\cdots COF} = V^{HO}(\alpha) + V^{HF}(\alpha + \pi)$$
$$= (V_1^{HO} - V_1^{HF}) \cos\alpha + (V_2^{HO} + V_2^{HF}) \cos 2\alpha + \ldots$$

The total interaction energy between the CH_2F and COF group is

$$V(\alpha) = V^{F\cdots COF}(\alpha) + V^{H\cdots COF}(\alpha + 2\pi/3) + V^{H\cdots COF}(\alpha - 2\pi/3) + \ldots$$
$$= [(V_1^{FO} - V_1^{FF}) - (V_1^{HO} - V_1^{HF})] \cos\alpha$$
$$+ [(V_2^{FO} + V_2^{FF}) - (V_2^{HO} + V_2^{HF})] \cos 2\alpha + \ldots$$

Apart from the additive constant this expression is equivalent to the experimentally determined potential function in the form,

$$V(\alpha) = V_0 + V_1 \cos \alpha + V_2 \cos 2\alpha + \dots$$

and therefore the corresponding Fourier coefficients may be equated.

The first three coefficients, as evaluated from a limited amount of experimental data, have the following approximate values (in $kJ\,mol^{-1}$) $V_1 = -3\cdot0$, $V_2 = 18\cdot1$ and $V_3 = 8\cdot2$. The large value of V_2 shows that the interactions between a fluorine atom and the COF group is considerably different from that between a hydrogen atom and the same group. The small magnitude of the V_1 term shows that the V^{FO} and V^{FF} interactions must be rather similar and that the V^{HF} and V^{HO} interactions must either be very small or nearly equal. Since oxygen and fluorine atoms have similar sizes and electronegativities such conclusions are not unreasonable. The V_3 term has the same order of magnitude as the barriers in CH_3COF, CF_3COF and other molecules containing CH_3 or CF_3 groups.

4.5 The Empirical Approach to the Computation of Conformational Energies and Barriers

An empirical method of computing conformational energies and barriers based on the concept of strain energy and using classical mechanics is frequently used in the conformational analysis of organic molecules.[8] From this point of view the stable conformers of a molecule represent minimum values of the strain energy while conformers representing maxima in the molecular potential energy surface correspond to maximum values of the strain energy. The strain energy (SE) may be expressed as a sum of four components which depend on different types of geometrical parameters,

$$SE = E_r(r) + E_\theta(\theta) + E_\alpha(\alpha) + E_{nb}(d) \tag{4.1}$$

where $E_r(r)$ is due to bond length (r) deformations,
$\quad E_\theta(\theta)$ to bond angle (θ) deformations,
$\quad E_\alpha(\alpha)$ to torsion (α) about bonds, and
$\quad E_{nb}(d)$ to repulsive interactions between non-bonded atoms or groups separated by distance (d).

The problem of computing the geometries of the stable conformers of a molecule amounts to minimizing the strain energy (equation (4.1)) with respect to the various geometrical parameters. In order to do this it is necessary to have a set of normal or strain free values for the geometrical parameters from which the magnitudes of the deformations may be measured. It is also necessary to be able to express the different components of the strain energy as functions of the relevant geometrical parameters. The total strain energy is then calculated from equation (4.1) using classical mechanics.

The strain energies due to bond length and bond angle deformations are usually expressed by simple harmonic oscillator equations,

$$E_r(r) = \sum k_{r_i} (r_i - r_i^0)^2$$
$$E_\theta(\theta) = \sum k_{\theta_i} (\theta_i - \theta_i^0)^2$$

$$(4.2)$$

where r_i^0 and θ_i^0 represent the normal or strain free values of the bond lengths and bond angles and the summations are over all the bond lengths or angles in the molecule. The torsional strain is usually represented by a sum of simple cosine potential functions and its contribution will be of the form,

$$E_\alpha(\alpha) = \tfrac{1}{2} \sum_j \sum_n V_{jn} \left[1 - \cos n(\alpha_j - \alpha_j^0) \right] \qquad (4.3)$$

where n is over the number of Fourier coefficients required to define the jth torsion and j is over the number of torsional degrees of freedom in the molecule. Usually only the dominant Fourier coefficient for each torsion is included in equation (4.3). The non-bonded interactions are usually expressed using either a Lennard–Jones potential,

$$V_{LJ}(d) = \frac{a}{d^{12}} - \frac{b}{d^6} \qquad (4.4(a))$$

or a Buckingham potential

$$V_B(d) = a' \, e^{-b'/d} - \frac{c'}{d^6} \qquad (4.4(b))$$

where a, b or a', b', c' are constants and d is the distance between two non-bonded atoms or groups. The force constants in the expressions for the bond length and bond angle deformations (equations (4.2)) and the Fourier coefficients for the torsional deformations (equation (4.3)) are generally average values taken from related molecules. The constants in the expressions for the non-bonded interactions (equations 4.4)) have in many cases been estimated by measurement of the physical properties of the atoms or groups cerned from thermodynamic data.

Calculations of this type are now performed as a matter of routine on large numbers of molecules by computational methods. The various methods differ basically in the level to which they approximate the strain energy and the techniques used to locate the energy minima. Sometimes the bond lengths, or both bond lengths and bond angles, are kept fixed at predetermined values, thus considerably reducing the number of parameters with respect to which the energy must be minimized. A simple illustration of the methods described in this section is given in Chapter 8.

4.6 Ab initio Calculations of Conformation Energies and Barrier Heights

From the discussion of Chapter 2 it will be appreciated that the computation of conformation energies and barrier heights involves calculating the electronic energy of the molecule for different nuclear configurations. The electronic energy may be expression as the sum of four terms,

$$W = T_e + V_{ee} + V_{nn} + V_{en} \qquad (4.5)$$

where T_e is the kinetic energy of the electrons,
V_{ee} represents electron–electron repulsion,
V_{nn} represents internuclear repulsion, and
V_{en} represents electron–nuclear attraction.

In this expression the first three terms make a positive contribution to the electronic energy and their sum is referred to as the repulsive term while the fourth term is negative and is referred to as the attractive term. In this context the terms attractive and repulsive refer to the sign of the contribution to the potential energy function governing the vibrational or conformational behaviour of the nuclei. In the case of a diatomic molecule the potential energy curve would be calculated by evaluating equation (4.5) for different values of the internuclear distance. In much the same way, the calculation of the barrier to internal rotation in ethane involves calculating the electronic energy at the minimum or staggered conformation and at the maximum or eclipsed conformation. The energy at these conformations should be minimized with respect to the internal molecular parameters, i.e. the C—H and C—C bond lengths and the HCH angle, in order to determine the barrier accurately. In the case of large molecules it is seldom possible to optimize the energy with respect to all of the variables and often the only internal parameter considered is the torsional angle for the bond around which the internal rotation is taking place.

4.6.1 Ab initio molecular orbital computations[9]

The calculation of the total electronic energy for a molecule requires solving the Schrödinger wave equation for the electrons moving in the potential field of the fixed nuclei,

$$H_{elec} \, \psi_{elec} = E_{elec} \, \psi_{elec}$$

and then adding the coulombic repulsions of the nuclei. Most computations are made using the self consistent field linear combination of atomic orbitals molecular orbital (SCF LCAO MO) technique and details may be found in textbooks on quantum chemistry. The most accurate computations of this type are intended to give wave functions and energies which are as close as possible to the so called Hartree–Fock limit, which represents the best that can be achieved with such wave functions and is itself an approximation. It is well known that energies calculated using Hartree–Fock wave functions are some 0·5–1 % greater than the true molecular energies. Since the barrier to internal rotation in ethane is only some 5×10^{-3} % of the electronic energy it may be argued that the origin of the barrier may not even be determined within the Hartree–Fock approximation. This conclusion is unduly pessimistic since there are good reasons for assuming that most of the differences between the Hartree–Fock and true energies are independent of molecular conformation.

The Hartree–Fock wave function underestimates the repulsions between electrons having opposite spins but occupying molecular orbitals localized in the same regions of space. This is usually called the electron correlation energy and would be expected to depend primarily on the number of electron pairs. The correlation energies of H_2 ($1·02 \times 10^2$ kJ mol^{-1}), He ($1·10 \times 10^2$ kJ mol^{-4}) and C^{4+} ($1·19 \times 10^2$ kJ mol^{-1}) each of which has one pair of electrons lend support to this suggestion. The correlation energy is therefore expected to be very similar for the different rotamers of a given molecule and this is thought to be one of the main reasons for the success of molecular orbital computations.

Table 4.3 shows the results of some molecular orbital computations of varying degrees of sophistication for ethane and it can be seen that all the

Table 4.3   Results of some molecular orbital calculations for ethane

	Energy of eclipsed form (Hartrees)[a]	Energy of staggered form (Hartrees)	Barrier (kJ mol^{-1})
Pitzer and Lipscomb	−78·98593	−78·99115	13·8
Pitzer	−79·09233	−79·09797	14·6
Clementi and Davies	−79·10247	−79·10824	15·1
Fink and Allen	−79·14377	−79·14778	10·5
Veillard	−79·23411	−79·23900	12·8
Experimental		−79·84	12·3

[a] 1 Hartree = $2·62 \times 10^3$ kJ mol^{-1}.

calculations give reasonably good agreement with the experimental barrier. All of the computations, except the last one, were made using the experimentally observed geometry of the staggered form of ethane assuming a simple rotation of the methyl groups. In the last computation relaxation of the internal parameters was permitted between the staggered and eclipsed positions and it was found that the C—C bond increased by 20 pm and the HCH angles decreased by 0·5° at the eclipsed position. Table 4.3 shows that reasonable values for the barrier height may be obtained with wave functions which give energies significantly higher than the Hartree–Fock limit. Table 4.4 gives a comparison of the observed and computed barriers for a number of other molecules and the generally good agreement is further confirmation that computations of this type may be used as a basis for discussing the origin of barriers.

Table 4.4   Comparison of computed barriers (*ab initio* molecular orbital) with experimental values for some selected molecules

	Computed barrier (kJ mol^{-1})	Experimental barrier (kJ mol^{-1})
Ethyl fluoride	10·9	13·9
Acetaldehyde	4·6	4·8
Methanol	4·4	4·5
Methyl silane	6·0	7·1
Hydrogen peroxide (*trans*)	34·7	29·3
Hydrogen peroxide (*cis*)	4·6	4·6

4.6.2 Energy component analysis of rotational barriers

One approach to the problem of the origin of barriers would be to study how the individual energy terms of equation (4.5) vary during the internal rotation. Unfortunately these terms are found to be rather sensitive to small changes in the wave function, which arise from differences in the chosen basis sets of orbitals, and to small differences in molecular geometry. For ethane it is not unusual to find computations where the contributions from the individual energy terms to the barrier have opposite signs even though the computed values of the total energy and barrier are very similar. In order to overcome these difficulties attempts have been made to partition the total energy into two components which like the barrier show invariance to the above factors.[10]

The most successful two-component division of the total energy has been to split it into an attractive term ($V_a = V_{en}$) and a repulsive term ($V_r = V_{ee} + V_{nn} + T_e$). Figure 4.6 shows the potential energy curve for the internal rotation in ethane separated into its attractive and repulsive components.

The variation of both the attractive (ΔV_a) and repulsive (ΔV_r) components with torsional angle is considerably larger than the barrier but they are out of phase with each other. The barrier is said to be repulsive dominant since ΔV_r is larger than ΔV_a. There is a general similarity between Fig. 4.6 and the curves for the interaction of two helium atoms. This is believed to show that the C—H bonds at either end of the ethane molecule are interacting in a manner analogous to the familiar repulsions that occur when two helium atoms are brought close to each other. The results of computations and analyses of this type suggest that the origin of the ethane barrier is not due to interactions between C—H bond dipoles or higher multipole interactions. Similarly the approach of Pauling which requires the participation of carbon 3d orbitals in the bonding does not appear to be justified since the molecular orbital calculations have been made without inclusion of these orbitals.

Energy component analyses of barrier calculations for acetaldehyde and some other molecules reveal that their barriers are dominated by the attractive component ΔV_a. This is contrary to the long held view that rotational barriers are produced by repulsive forces. The attractive or repulsive nature of the barriers for several molecules for which accurate computations have been carried out is given in Table 4.5. This has made it possible to draw a number of important conclusions regarding the roles of X—H bond pairs and lone pairs of electrons in determining the nature of barriers.

Although the partitioning of the barrier energy into attractive and repulsive components has met with considerable success there are indications that in some cases it is not independent of the choice of basis set or the molecular

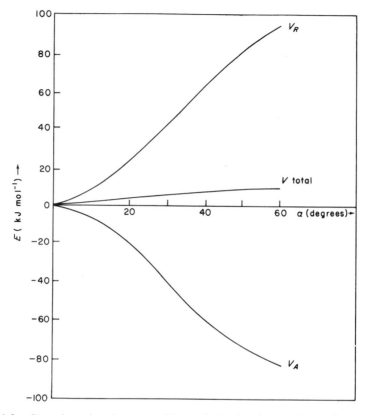

Fig. 4.6   Showing the decomposition of the barrier to internal rotation in ethane into its attractive and repulsive components.

Table 4.5   The dominant component (i.e. attractive or repulsive) found for the internal rotation barrier for some selected molecules

	Nature of barrier
Ethane	repulsive
Methanol	repulsive
Methylamine	repulsive
Ethyl fluoride	repulsive
Acetaldehyde	attractive
Hydroxylamine	attractive
Nitrosomethane	attractive

geometry used in the computations. The *cis* barrier in hydrogen peroxide was found to be attractive dominant in a number of computations and repulsive dominant in others.

This type of analysis has also been applied to the inversion of ammonia and to some other simple amines. Internal motions of this type which involve a change in hybridization are referred to as "first order", while a process such as the internal rotation in ethane which involves no drastic change in chemical bonding is referred to as "second order". It is more difficult to obtain accurate results for first order processes.

In concluding this section it must be mentioned that some radically different schemes for partitioning the total energy have been suggested and applied to a few molecules. Any such scheme which proves to be relatively insensitive to the size of basis set and small changes in molecular geometry may well replace the attractive and repulsive partitioning of the total energy as a means of investigating the origin of barriers.

4.6.3 Charge distribution analysis of rotational barriers

Another approach to the origin of barriers, perhaps with a greater intuitive appeal to chemists, is to examine the electron charge distributions in different rotamers. This may be done in a variety of ways including plotting the difference in charge distributions between two rotamers.[11] In ethane this procedure has shown that there is a small increase in the electron density behind the hydrogens in the eclipsed form. This is consistent with increased repulsions between the eclipsing hydrogen atoms in this rotamer. In the equilibrium conformation of acetaldehyde there is an indication of the formation of a weak covalent or hydrogen bond between the oxygen atom and the hydrogen atom it eclipses. Electron population analyses have also been used to investigate the charge distributions in different rotamers. In ethane a number of calculations indicate that there is a slight increase in the C—C overlap population of the staggered rotamer compared to the eclipsed rotamer suggesting a slightly stronger C—C bond in the more stable conformer.

4.7 Conclusion

This chapter has been concerned mainly with the computation of conformational energies and barrier heights by semi-empirical and *ab initio* methods. Semi-empirical calculations have now reached the stage where they can be applied to quite large molecules and the results are often very useful in the interpretation of experimental data where conformational behaviour is believed to have important consequences. *Ab initio* calculations on small molecules now give rotational barriers in very good agreement with the experimental values. Interpretation of the results of these computations has led to a great deal of insight into the origin of barriers and it has been possible to show that a number of the earlier suggestions are not tenable. However, it would be too optimistic to suggest even now that no major problems remain in the calculation of conformational energies.

Further Reading

Some excellent reviews of the origin of potential barriers and the methods used to calculate them are available. The interested reader is directed to one of the following articles or to refs 1 and 9, where the subject matter of Chapter 4 is dealt with in greater detail.

1. J. P. Lowe. *In* "Progress in Physical Organic Chemistry" (A. Streitweizer and R. Taft, Eds), p. 1. Interscience, Chichester and New York, 1968.
2. E. B. Wilson. *Chemical Soc. Reviews*, **1**, 293 (1972).

References

1. E. B. Wilson. *In* "Advances in Chemical Physics" (I. Prigogine, Ed.), Vol. 2. p. 367. Interscience, Chichester and New York, 1959.
2. E. B. Wilson. *Chemical Soc. Reviews* **1**, 293 (1972).
3. G. I. L. Jones and N. L. Owen. *J. Mol. Structure*, **18**, 1 (1973).
4. E. Hirota, R. Sugisaki, C. J. Nielsen and G. O. Sørensen. *J. Mol. Spectroscopy*, **49**, 251 (1974).
5. K.-M. Marstokk and M. Møllendal. *J. Mol. Structure*, **18**, 247 (1973).
6. J. M. Riveros and E. B. Wilson. *J. Chem. Phys.* **46**, 4605 (1967).
7. E. Saegebarth and E. B. Wilson. *J. Chem. Phys.* **46**, 3088 (1967).
8. J. E. Williams, P. J. Stang and P. von R. Schleyer. *Annual Review of Physical Chemistry*, **19**, 531 (1968).
9. A. Veillard. Ab initio calculations of barrier heights, *in* "Internal Rotation in Molecules" (W. J. Orville-Thomas, Ed.), Chap. 11. Wiley, Chichester and New York, 1974.
10. L. C. Allen. *Chem. Phys. Letters*, **2**, 597 (1968).
11. W. L. Jorgensen and L. C. Allen. *J. Amer. Chem. Soc.* **27**, 55 (1971).

5
Internal Rotation of
Symmetric Groups

5.1 Introduction

Hindered internal rotation of a symmetric group, such as CH_3, about a single covalent bond, gives rise to a periodic variation in the total potential energy of a molecule. The shape of the potential curve for a threefold symmetric internal rotor as a function of angle of internal rotation has been shown in Fig. 1.6, and a mathematical formulation of the curve is given in Chapter 2. The forces involved in generating barriers to internal rotation in molecules have also been discussed earlier, in Chapter 4. Although many different techniques may be used to determine the heights of potential barriers, most of the finer details of the potential energy profiles and most of the accurate barrier heights for methyl groups have been derived from studies of pure rotation spectra (in the microwave region of the spectrum). In this chapter the theoretical background involved in the main spectroscopic methods of measuring potential barriers will be reviewed and the significance of some of the results discussed.

5.2 Ethane

The ethane molecule which comprises two methyl groups linked by a covalent bond, represents the prototype for the phenomenon of internal rotation in molecules. What is so intriguing about ethane, however, is the fact that the molecule is not very amenable to direct study of the internal rotation. There is no permanent dipole and so no rotational spectrum; also, because of the molecular symmetry, the torsion does not involve a change of dipole or a change of polarizability, and so is inactive in both infrared and Raman spectra.

Consequently for many years the barrier height to internal rotation in ethane was less precisely determined than barriers in many, much more complex, molecules. Because the torsion in ethane is so inaccessible to spectroscopic investigation, most of the early barrier determinations were based on entropy and low temperature heat capacity measurements. It reflects well on the care and precision of the early pioneers in this field that the currently accepted value of the barrier height for ethane (12.25 ± 0.10 kJ mol^{-1}) falls close to the best thermodynamic measurement obtained in 1951 (12.03 ± 0.52 kJ mol^{-1}), and lies within the limits of the estimated experimental uncertainties quoted for the latter.

Early infrared and Raman analyses enabled estimates to be made of the frequency of the inactive torsional mode from the appearance of combination bands, and in 1949 a careful high resolution study of these combination bands in the gas phase infrared spectrum showed convincingly that the stable conformation of ethane has a staggered arrangement of C—H bonds[1] (i.e. ethane has point symmetry elements corresponding to the D_{3d} group). The development of inelastic neutron scattering methods has led to an independent estimate of the torsional frequency but the most accurate currently accepted value is derived from the infrared spectrum of ethane gas at high pressure.[2] Under high pressure conditions the torsional mode becomes weakly active owing to slight molecular distortions brought about by intermolecular collisions.

It is interesting to note that the microwave spectra of the ground and first excited torsional state of CH_3—CD_3 have been observed[3]. The reported relative intensities of rotational lines originating in these states are consistent with the currently accepted value for V_3 for ethane. The very small polarity difference between C—H and C—D bonds results in the CH_3CD_3 molecule possessing a small but finite permanent dipole moment enabling the pure rotation spectrum to be observed.

5.3 Torsional Energy Levels

The potential energy function for an internal rotor with threefold symmetry (e.g. ethane and all monosubstituted ethanes) has three identical minima for each 360° rotation of the methyl group, and when the potential barrier between each minimum is infinitely high, all energy levels within the minima are triply degenerate and the system comprises three localized wavefunctions, ψ_I, ψ_{II} and ψ_{III}, (Fig. 5.1). However, as the barrier is reduced, the possibility of quantum mechanical tunnelling increases and the degeneracy is partially

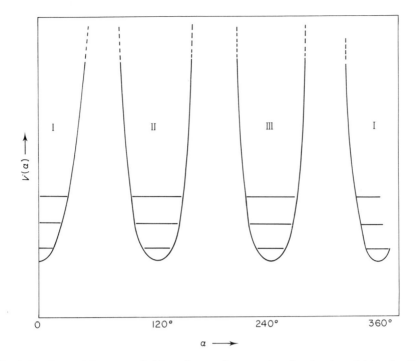

Fig. 5.1   Part of the potential function and energy levels of a threefold potential with an infinitely high barrier.

lifted giving a non-degenerate A state and a doubly degenerate E state. The allowed wavefunctions for such a system with three identical minima are,

$$\psi_A = \frac{1}{\sqrt{3}}(\psi_I + \psi_{II} + \psi_{III})$$

$$\psi_{E_1} = \frac{1}{\sqrt{3}}(\psi_I - \psi_{II} + \psi_{III})$$

$$\psi_{E_2} = \frac{1}{\sqrt{3}}(\psi_I + \psi_{II} - \psi_{III})$$

5.3.1 Symmetry considerations

Some insight into the torsional energy levels associated with a threefold symmetric potential may be obtained by considering the symmetry of a CH_3 group. In terms of point symmetry the rotations of a methyl group are incorporated in the symmetry group C_3 (Table 5.1). This is a point group with

Table 5.1 C_3 point symmetry group character table, showing symmetry elements and irreducible representations

Symmetry species	I	$C_3{}^a$	C_3^2
A	1	1	1
$E_1\rbrace E$	1	ε	ε^*
E_2	1	ε^*	ε

[a] C_3 represents a rotation by $2\pi/3$ about the symmetry axis: $\varepsilon = e^{2\pi i/3}$ and ε^* represents the corresponding complex conjugate; $\varepsilon\varepsilon^* = 1, (\varepsilon)^2 = \varepsilon^*, (\varepsilon^*)^2 = \varepsilon$.

three operations, the identity (I), rotation by 120° (C_3) and by 240° (C_3^2) about the symmetry axis.

A comprehensive group theoretical discussion of the symmetry of molecules with hindered internal rotation is of necessity more complicated than that for rigid molecules since the overall molecular symmetry may be affected by the internal motion[4]. For the purpose of this discussion, however, only the symmetry properties of the internal top need be considered. For any molecule with an asymmetric framework containing a threefold symmetric internal rotor the symmetry group C_3 is a sub-group of the overall molecular symmetry group. Therefore the total molecular Hamiltonian must be invariant to operations within the C_3 group when they are carried out only on the methyl group. The character table of this group (Table 5.1) shows that there are three irreducible representations, two of which (E_1 and E_2) are degenerate. The molecular dipole moment is not affected by any of the operations of the group and must therefore transform with symmetry A. For any transition to be allowed, the transition moment $\langle \psi_i | \mu | \psi_j \rangle$ must be finite. Consequently, since μ is symmetric with respect to operations within the C_3 group, all allowed transitions must connect energy states (i and j) of the same symmetry, i.e. $A \leftrightarrow A$; $E \leftrightarrow E$; $A \nleftrightarrow E$. These selection rules are generally valid for all molecules with internal rotation, although other specific symmetry connections may be derived for molecules with higher molecular symmetry.

The torsional wavefunctions of a methyl group are generally labelled thus: $\psi_{v,\sigma}$ where v represents the torsional quantum number and σ defines the symmetry of the torsional state; $\sigma = +1$ and $\sigma = -1$ corresponding to the two degenerate states (E_1 and E_2) and $\sigma = 0$ to the non-degenerate (A) state. The relationship between the energy levels of a hindered internal rotor and those of a free rotator, and the correspondence between the respective quantum labels "v" and "m" is illustrated in Fig. 5.2. The label v is particularly

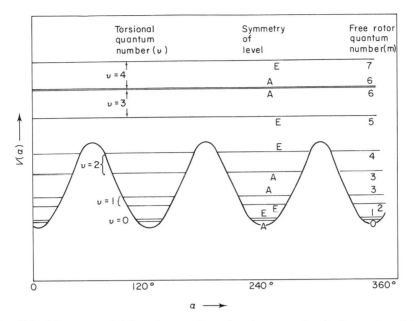

Fig. 5.2 The potential function and torsional energy levels for a threefold symmetric rotor, showing the correlation between the hindered internal rotation quantum number (v) and the free internal rotation quantum number (m).

useful within the potential minima, and m is a "good" quantum number for free internal rotation.

The effect of quantum mechanical tunnelling through the potential barrier is to split the triply degenerate energy states into *two* levels, A and E. This splitting increases as the levels approach the top of the barrier and as tunnelling becomes easier. It may be noted in Fig. 5.2 that the symmetry of the levels alternates within the potential wells.

5.3.2 Spin statistical weights and absorption line intensities

Because tunnelling has the effect of splitting the energy levels into A and E sub-levels many microwave transitions appear as doublets, each component corresponding to a transition between substates of the same symmetry. To determine the relative intensities of the A and E component lines in a rotational transition the symmetry of the wavefunctions must be considered. Only the torsional and nuclear spin wavefunctions are important in this context since the electronic and rotational wavefunctions have completely symmetric properties with respect to the internal rotation. Suitable combinations of torsional and nuclear spin wave functions are governed by the fact that the overall wavefunction must be antisymmetric with respect to interchange of any one pair of protons (protons being fermions). For a CH_3 group there are eight nuclear spin functions which must be combined with the torsional wavefunctions (Fig. 5.3).

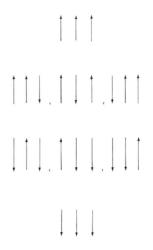

Fig. 5.3   Representation of the eight nuclear spin functions of the protons of a CH_3 group.

These spin functions form a reducible representation of the C_3 group, with characters that are determined by the number of functions left unaltered by any operation within the group $[\chi(E) = 8;\ \chi(C_3) = 2;\ \chi(C_3^2) = 2]$. This reducible representation may be shown through standard group theoretical procedures to comprise the following irreducible representations of the C_3 group $(4A + 2E_1 + 2E_2)$.

The symmetry of the product wavefunctions $\psi_{\text{torsion}}\,\psi_{\text{spin}}$ is determined for the A and E torsional states by direct multiplication of the appropriate characters, with the following results,

$$A \times (4A + 2E_1 + 2E_2) = 4A + 2E_1 + 2E_2$$

$$(E_1 + E_2) \times (4A + 2E_1 + 2E_2) = 4A + 6E_1 + 6E_2$$

Since the overall wavefunction must be antisymmetric with respect to interchange of any two protons, and since any operation of the C_3 group interchanges an even number of pairs of protons, it follows that the total wavefunction must be totally symmetric (i.e. belong to the symmetry species A). The nuclear spin weight of a level is thus given by the weight of the representation A in the product wavefunctions. Examination of the expressions given above shows that the non-degenerate and the degenerate levels have equal weights (i.e. 4). Consequently the A and E component lines of a microwave transition have equal intensity.

For a CD_3 group the situation is more complicated, since there are twenty-seven nuclear spin states present owing to the fact that deuterium has three spin states $(+1, 0, -1)$. A similar analysis of the symmetry properties of the product wave function $(\psi_{torsion} \psi_{spin})$ shows that the ratio of the $A:E$ line intensities for a CD_3 group will be $11:16$. This applies both to molecules with planar and asymmetric frameworks. This intensity difference between A and E components arising from CH_3 and CD_3 groups can prove useful in the analysis of spectra of molecules with internal rotation.

5.4 The Calculation of Potential Barriers

5.4.1 Torsional frequency method

The most direct method of estimating the barrier for a symmetric group is through measurement of the fundamental (i.e. $v = 1 \leftarrow v = 0$) of the torsional mode. This is most conveniently done by directly observing the absorption in the far-infrared or Raman spectrum, or more indirectly by neutron scattering, or from intensity measurements in the microwave spectrum.

For the purpose of most barrier calculations, a potential with a periodicity "n", may be adequately expressed as a simple cosine function of the torsional angle (see Chapter 2),

$$V(\alpha) = \frac{V_n}{2}(1 - \cos n\alpha) \tag{5.1}$$

where only the first term in the general expression has been retained. This equation may be applied to internal rotation of groups with values of n other than 3, for example nitrobenzene (where $n = 2$). However, for symmetric top internal rotors $n \geqslant 3$, and the overall rotational energy is independent of the angle of internal rotation (α). This independence of the overall inertial parameters on the internal motion makes treatment of internal rotation for symmetric top rotors particularly simple.

Equation (5.1) may be used in one of several different ways depending on the height of the barrier and on the accuracy required.

5.4.2 High barriers

For very high barriers, the torsional mode approximates to small amplitude oscillations about the potential minima.

Equation (5.1) may be rewritten as,

$$V(\alpha) = V_n \sin^2\left(\frac{n\alpha}{2}\right) \tag{5.2}$$

For low amplitude oscillations, $\sin \alpha \simeq \alpha$ (in radians), thus,

$$V(\alpha) = V_n \left(\frac{n\alpha}{2}\right)^2 \tag{5.3}$$

When this expression for the potential energy is substituted into the Schrödinger equation for a one dimensional oscillator we have,

$$F\frac{d^2\psi(\alpha)}{d\alpha^2} + \left[E - \frac{V_n n^2 \alpha^2}{4}\right]\psi(\alpha) = 0 \tag{5.4}$$

Here, F is the reduced rotational constant for the system, $F = h^2/8\pi^2 I_r$, where I_r (the reduced moment of inertia) $= I_\alpha(1 - \Sigma\lambda_i^2 I_\alpha/I_i)$, (see Section 5.5.1). Equation (5.4) has the same overall form as that of a harmonic oscillator, and has solutions of similar form; viz.

$$E = (v + \tfrac{1}{2})\,n\sqrt{V_n F} \tag{5.5}$$

The frequency of such a harmonic oscillator (c.f. $E = hv(v + \tfrac{1}{2})$) is given by,

$$v = \frac{n}{h}\sqrt{V_n F} \tag{5.6}$$

$$\therefore v = \frac{1}{2\pi}\sqrt{\frac{V_n n^2}{2I_r}} \tag{5.7}$$

Thus the potential barrier is proportional to the square of the torsional frequency, e.g. for a methyl group, $n = 3$, equation (5.7) becomes,

$$V_3 = \tfrac{8}{9}\pi^2 v^2 I_r$$

5.4.3 Intermediate barriers

The insertion of the general expression for a cosine potential function into the Schrödinger equation results in the following equation,

$$F \frac{d^2 \psi(\alpha)}{d\alpha^2} + \left[E - \frac{V_n}{2} (1 - \cos n\alpha) \right] \psi(\alpha) = 0 \qquad (5.8)$$

Equation (5.8) is very similar in form to a well-known differential equation known as the Mathieu equation,

$$\frac{d^2 y}{dx^2} + (b - s \cos^2 x) y = 0 \qquad (5.9)$$

The equivalence of these two equations can be made complete by the following transformations,

$$\psi(\alpha) = y \qquad (5.10a)$$

$$(n\alpha \mid \pi) = 2x \qquad (5.10b)$$

$$V_n = \frac{n^2 F s}{4} \qquad (5.10c)$$

$$E = \frac{n^2 F b}{4} \qquad (5.10d)$$

where b is an eigenvalue of the Mathieu equation, s is a parameter known as the reduced barrier, and where $y = f(x)$.

The boundary conditions associated with the Mathieu equation imply periodic solutions of the equation with period $n\pi$ in x (i.e. period 2π in α). Thus the Mathieu functions periodic in π will be applicable for any value of n, but those periodic in 2π or 3π, etc., will also be required when n is a multiple of 2 or 3, etc. This leads to a number of distinct sub-levels for each torsional level, there being $(n + 1)/2$ sub-states where n is odd, and $(n + 2)/2$ where n is even (e.g. for $n = 3$, we have two states, $\sigma = 0$ and $\sigma = \pm 1$). The torsional energy levels (equation (5.10d)) correctly subscripted can thus be written as,

$$E_{v\sigma} = \frac{n^2}{4} F b_{v\sigma} \qquad (5.11)$$

Since spectroscopy measures differences in energy levels and an absorption line corresponds to an energy change $E_2 - E_1$ equation (5.11) may be written,

$$\Delta E_{v\sigma} = \frac{n^2}{4} F \Delta b_{v\sigma} \qquad (5.12)$$

If sufficient structural information is available to determine F for a molecule, equation (5.12) may be used to evaluate the parameter Δb_v. The relationship between Δb_v and the dimensionless reduced barrier s may be found conveniently from published tables of solutions to the Mathieu equation, and the potential barrier V_n calculated from the reduced barrier s by means of equation (5.10c).

5.4.4 Assessment of the torsional frequency method

The calculation of barriers from torsional frequencies hinges on the assumption that the torsional mode of a molecule exhibiting internal rotation may be considered separately from all other vibrations of the molecule. For simple systems, such as substituted ethanes, this is a reasonable assumption, but for more complex and flexible molecules with several internal axes of rotation, the possibility of the torsion being coupled with one or more of the other vibrations is considerable. The coupling together of vibrations can occur through both potential and kinetic energy terms, although in the case of torsions interacting with other vibrations the former predominates (see Chapter 2).

Agreement between the harmonic oscillator and Mathieu equation methods of calculating barriers from torsional frequencies is reasonably good for fairly high barriers (i.e. $V_3 > 16 \text{ kJ mol}^{-1}$), but the former method generally predicts lower values for the barrier. Improvement of the agreement between the two calculations may be achieved if further terms in the expansion of $\sin \alpha$ (equation (5.3)) are included as a perturbation in equation (5.4).

5.5 Splittings Method

Most potential barriers hindering the internal rotation of symmetric top groups have been determined from analyses of pure rotational spectral line frequencies. The basis of the method lies in the fact that the overall rotational motion of a molecule is coupled to the internal rotation. For asymmetric molecules the coupling differs for the two torsional sub-levels (A and E), and is a function of the rotational quantum numbers. Consequently a rotational transition may be split into two components corresponding to transitions involving the A and E sub-levels. The extent of the splitting is a sensitive function of the potential barrier to internal rotation. For symmetric top molecules, (e.g. CH_3CF_3) the effect of coupling between the internal and overall rotational motions cannot generally be observed in the microwave spectrum since the interaction is independent of the rotational quantum number J and the upper and lower rotational energy levels are affected equally (Fig. 5.4). The determination of barriers for symmetric top molecules is discussed further in Section 5.7.

The most convenient molecular model for the treatment of internal rotation consists of a rigid framework, which in the general case is asymmetric, with a symmetric top attached to it. The assumption is made that the hindered internal rotation may be treated separately from the other normal vibrations of the molecule, an assumption which has been proved to be satisfactory provided that the torsion is much lower in frequency than the other vibrations.

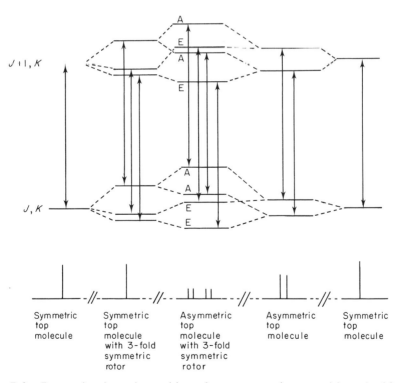

Fig. 5.4   Energy levels and transitions for a symmetric top, with and without a threefold rotor, and for an asymmetric top with and without a threefold rotor, showing the correlation between the energy levels.

Two main approaches have been used for the internal rotation problem. The methods differ in that the Hamiltonian operator is expressed with reference to different inertial axes systems. The principal axis method (PAM) utilizes the three principal axes of the molecule, and the axis of internal rotation coincides with the symmetry axis of the top, which in general is not coincident with any one of the principal axes. In the internal axis method (IAM) on the other hand, one of the coordinate axes of the overall molecular system is arranged to coincide with or to be parallel with the axis of the internal rotor. The other two axes are chosen within the molecular framework such that the origin lies at the centre of mass. An advantage of the IAM lies in the fact that the interaction energy terms between the torsional motion and the overall rotation of the molecule are minimized compared to those involved in the PAM. On the other hand the IAM suffers from the disadvantage of a more complicated Hamiltonian since terms involving products of inertia are introduced. Also with the internal axis system, more complicated, non-periodic Mathieu functions have to be used.

Either approach can be utilized successfully for most molecules with internal rotation, although where the symmetric top is heavier than the asymmetric framework (e.g. CH_3OH, CF_3CHO), the IAM is more suitable. The PAM probably represents the simplest method for calculating internal rotation barriers for a typical molecule since the required perturbation coefficients for all barrier values are available in tabulated form.

In outline, the calculation of barriers by the splittings method is as follows,

(a) The splittings between the rotational lines are measured accurately.
(b) The splittings between the absorption line doublets or the differences between the rotational constants for the two sets of lines, are related to dimensionless coefficients. At this juncture the structure of the molecule (in the form of the parameter F) enters into the calculation (equation (5.12)).
(c) The coefficients are related to the reduced barrier (s).
(d) The reduced barrier s is a simple function of the true barrier height and of the parameter F, (equation (5.10c)).

5.5.1 Principal axis method

If we consider only the kinetic energy of a molecule, then we have, using matrix notation,

$$2T = \omega^+ I \omega \tag{5.13}$$

where T represents the kinetic energy, $\mathbf{I}$ is the inertial tensor and $\boldsymbol{\omega}$ is the angular velocity vector. For symmetric top internal rotation $\mathbf{I}$ and $\boldsymbol{\omega}$ are defined as follows,

$$\boldsymbol{\omega} = \begin{bmatrix} \omega_a \\ \omega_b \\ \omega_c \\ \dot{\alpha} \end{bmatrix} ; \quad \mathbf{I} = \begin{bmatrix} I_a & 0 & 0 & (\lambda_a I_\alpha) \\ 0 & I_b & 0 & (\lambda_b I_\alpha) \\ 0 & 0 & I_c & (\lambda_c I_\alpha) \\ (\lambda_a I_\alpha) & (\lambda_b I_\alpha) & (\lambda_c I_\alpha) & I_\alpha \end{bmatrix}$$

$\dot{\alpha}$ is the angular velocity of the top relative to the frame; I_i is the ith principal moment of inertia; I_α is the moment of inertia of the symmetric top about its symmetry axis, and λ_i is the directional cosine between the internal rotor axis and the ith principal axis, (Fig. 5.5).

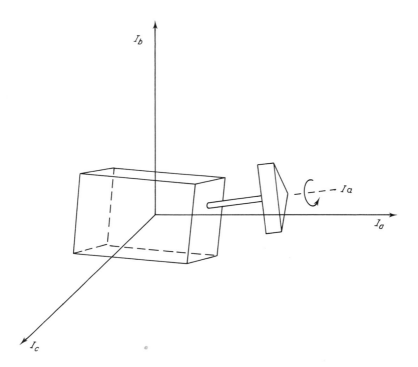

Fig. 5.5   Showing a molecule with a symmetric top in the Principle Axis System.

Thus, if we separate out the pure overall rotation terms, we have,

$$2T = \sum_i I_i \omega_i^2 + 2I_\alpha \dot{\alpha} \sum_i \lambda_i \omega_i + I_\alpha \dot{\alpha}^2 \tag{5.14}$$

(where $i = a, b, c$)

To derive the quantum mechanical Hamiltonian the kinetic energy has to be expressed in terms of angular momenta, i.e.

$$P_i = \frac{\partial T}{\partial \omega_i} \quad \text{and} \quad p = \frac{\partial T}{\partial \dot{\alpha}}$$

Thus,

$$P_i = I_i \omega_i + \lambda_i I_\alpha \dot{\alpha} \tag{5.15}$$

and

$$p = I_\alpha \dot{\alpha} + I_\alpha \sum_i \lambda_i \omega_i \tag{5.16}$$

It can be seen from the above equations that both P_i and p contain contributions from both overall and internal motions. They can, however, be combined as follows to represent a "relative" angular momentum $(p - \mathscr{P})$ which depends only $\dot{\alpha}$. Let

$$\mathscr{P} = \sum_i \frac{\lambda_i I_\alpha}{I_i} P_i \tag{5.17}$$

then

$$p - \mathscr{P} = I_\alpha \dot{\alpha} \left[1 - \sum_i \frac{\lambda_i^2 I_\alpha}{I_i} \right] \tag{5.18}$$

or

$$p - \mathscr{P} = r I_\alpha \dot{\alpha} \tag{5.19}$$

where

$$r = 1 - \sum_i \frac{\lambda_i^2 I_\alpha}{I_i} \tag{5.20}$$

From equations (5.14) and (5.15),

$$2T - \sum_i \frac{P_i^2}{I_i} = r I_\alpha \dot{\alpha}^2 \tag{5.21}$$

Thus from equations (5.19) and (5.21), the kinetic energy may be conveniently expressed as,

$$2T = \sum_i \frac{P_i^2}{I_i} + \frac{(p - \mathscr{P})^2}{rI_\alpha} \qquad (5.22)$$

The total Hamiltonian for the molecular system can now be written as,

$$H = H_r + F(p - \mathscr{P})^2 + V(\alpha) \qquad (5.23)$$

where H_r is the usual rigid rotor Hamiltonian $(AP_a^2 + BP_b^2 + CP_c^2)$ arising from the first term of equation (5.22), here A, B and C are rotational constants; F is the reduced rotational constant for the internal rotation $(F = h^2/8\pi^2 rI_\alpha)$; and $V(\alpha)$ represents the potential energy associated with the internal rotation. Therefore we can express the Hamiltonian as a sum of terms,

$$H = H_R + H_T + H_{TR}$$

where

$$\left.\begin{array}{l} H_R = H_r + F\mathscr{P}^2 \\[6pt] H_T = Fp^2 + V(\alpha) \\[6pt] H_{TR} = -2FQp \end{array}\right\} \qquad (5.24)$$

The term $F\mathscr{P}^2$ can be incorporated into the rigid rotor Hamiltonian thus slightly modifying the rotational constants. The expression $Fp^2 + V(\alpha)$ is equivalent to the torsional Hamiltonian given in equation (5.8), where the kinetic energy appears in its quantum mechanical form. The internal and overall rotational motions could be treated completely separately were it not for the interaction terms $-2F\mathscr{P}p$. For relatively high barriers (when the torsional energy level separations are considerably larger than the rotational energy level separations) these interaction terms may be treated as a perturbation on the Hamiltonian $(H_R + H_T)$.

The usual method of dealing with the H_{TR} terms is to consider the product wavefunction $\psi_R\psi_T$ as a basis in which both H_R and H_T are diagonal and H_{TR} only contributes non-diagonal elements in the torsional quantum number (v). By using a Van Vleck transformation (see Chapter 2) the off-diagonal terms of the energy matrix are reduced and the diagonal terms modified through inclusion of the diagonal products of the transformation. Such a transformation may be applied successively until the off-diagonal terms

E

become negligible. This results in an effective Hamiltonian for each torsional state. Thus,

$$H_{v\sigma} = H_R + FW_{v\sigma} \tag{5.25}$$

where the term $W_{v\sigma}$ includes all the correction terms including those which have been introduced by the Van Vleck transformation. These can be expressed as a power series in the reduced angular momentum $\mathscr{P}$,

$$W_{v\sigma} = \sum_n W_{v\sigma}^{(n)} \mathscr{P}^n \tag{5.26}$$

Since $\mathscr{P}$ is a function of the moment of inertia of the top relative to the framework (equation (5.17)), for light tops attached to heavy frameworks the above power series converge rapidly and only the first few terms need be considered.

The first term of this perturbation series ($n = 0$), describes the pure torsional energy of the system,

$$E_{v\sigma} = FW_{v\sigma}^{(0)} \tag{5.27}$$

and represents an eigenvalue of the Mathieu equation (c.f. equation (5.10d)), therefore for a methyl group,

$$W_{v\sigma}^{(0)} = \tfrac{9}{4}b_{v\sigma}$$

For very large barriers ($V_n \to \infty$) only this first term ($n = 0$) need be considered and the problem reduces to that of a harmonic oscillator.

The second term (which is the first order $n = 1$ contribution) of the series is $-2F\mathscr{P}p$, so that,

$$W_{v\sigma}^{(1)} = -2p$$

This is finite only for the E state levels ($\sigma = \pm 1$), and for fairly high barriers ($V_3 > 4 \text{ kJ mol}^{-1}$) it is not very significant even for the E energy levels. Where it does contribute, the E levels deviate a little from a rigid rotor pattern. (All higher order "odd" terms, $n = 3, 5$ etc., are also zero for A levels, since by symmetry arguments $\langle A|p^n|A\rangle = 0$ for all odd values of n).

The next term (second order $n = 2$ contribution) of the series differs in magnitude and often in sign for the A and E sub-levels of a given torsional state, thus resulting in slightly different Hamiltonians H_{vA} and H_{vE} with corresponding different spectral patterns.

All these perturbation coefficients $W_{v\sigma}^{(n)}$ are dimensionless (depending only on the ratio V_3/F) and have been tabulated for various values of n and v for a wide range of the reduced barrier s ($s = 4V_3/9F$, see equation (5.10c)), and are readily available in the literature. For molecules with low barriers, higher order terms in the series become increasingly important.

For relatively high barriers ($V_3 > 4$kJ mol^{-1}) where the odd order coefficients for the E levels are negligible, both A and E states follow a pseudo-rigid rotor pattern and the rotational constants take the form,

$$A_{v\sigma} = A + F\rho_a^2 W_{v\sigma}^{(2)} \tag{5.28}$$

where

$$\rho_a = \frac{\lambda_a I_a}{I_a}$$

The height of the potential barrier may be obtained directly from the differences between the A and E rotational constants, for example,

$$\Delta A = A_{vA} - A_{vE}$$
$$= F\rho_a^2 [W_{vA}^{(2)} - W_{vE}^{(2)}] \tag{5.29}$$

Values of $W_{vA}^{(2)}$ and $W_{vE}^{(2)}$ are available in the literature for various ranges of the reduced barrier s.

When the differences in the A and E rotational constants are too small to be useful for determining the barrier, a value may be obtained by measuring the separation between individual transitions. The splitting of the energy levels can be expressed as a function of the rotational constants as follows,

$$\Delta E = E_A - E_E$$
$$= \left(\frac{\partial E}{\partial A}\right)\Delta A + \left(\frac{\partial E}{\partial B}\right)\Delta B + \left(\frac{\partial E}{\partial C}\right)\Delta C \tag{5.30}$$

For a rigid rotor the rotational energy may be expressed in terms of the angular momenta as,

$$E = A\langle P_a^2 \rangle + B\langle P_b^2 \rangle + C\langle P_c^2 \rangle \tag{5.31}$$

and thus,

$$\frac{\partial E}{\partial A} = \langle P_a^2 \rangle$$

$$\frac{\partial E}{\partial B} = \langle P_b^2 \rangle \qquad (5.32)$$

$$\frac{\partial E}{\partial C} = \langle P_c^2 \rangle$$

The average values of the quadratic angular momenta can be obtained from tables of asymmetric rotor energy levels.

5.5.2 Internal axis method

When first developed the internal axis method was largely restricted to symmetric top molecules or to asymmetric top molecules with a plane of symmetry. For the more general case of a completely asymmetric framework, the approach was limited to the effects of internal rotation on low J energy levels owing to the complexity of the Hamiltonian. One of the coordinate axes of the molecule is chosen to be coincident with or to lie parallel to the axis of internal rotation of the top (hence the name Internal Axis Method) while the other two axes, passing through the centre of gravity, are fixed arbitrarily in the framework (Fig. 5.6). In this coordinate system the interaction terms between internal and overall rotation are much smaller than those arising in the PAM.

The overall Hamiltonian, however, is complicated by the presence of products of inertia terms since the coordinate axes are not the principal axes. For example, the Hamiltonian for a molecule with a plane of symmetry (xz plane), and with a methyl group lying in the symmetry plane has the form,

$$H = AP_y^2 + BP_x^2 + CP_z^2 + D(P_xP_z + P_zP_x) - 2DP_x p$$
$$- 2CP_z p + Fp^2 + V(\alpha)$$

where A, B, C and D are functions of the moments of inertia. Coordinate transformations may be applied to minimize the coupling terms between the top and framework and to ensure that there is no change of angular momentum about the top axis. More recent IAM treatments have been derived for molecules with completely asymmetric frameworks.[5]

The relative angular momentum of a molecule with a rotating symmetric

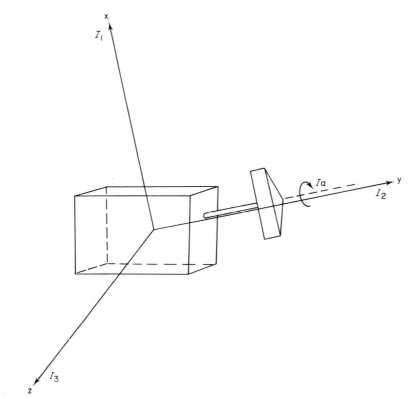

Fig. 5.6   Showing a molecule with a symmetric top in the Internal Axis System.

top is given by,

$$\mathscr{P} = \sum_i \frac{\lambda_i I_\alpha}{I_i} \, P_i$$

$$= \sum_i \rho_i P_i \tag{5.33}$$

Using equation (5.33), the molecular Hamiltonian has the same form as equation (5.23),

$$H = H_r + F(p - \rho P)^2 + V(\alpha) \tag{5.34}$$

Here ρ defines the direction of an axis, fixed in the framework of the molecule. This axis connects the centre of mass of the internal rotor to the centre of mass of the whole molecule.

The Hamiltonian (5.34) divides naturally into the rigid rotor energy (H_r)

and the internal rotation energy $(F(p - \rho P)^2 + V(\alpha))$, although the two parts are obviously connected through the appearance of the total angular momentum (P) in both. The internal rotation equation may be solved by transforming the torsional terms into a Mathieu differential equation (see, for example, Section 4.3, equation (5.10), but with the symmetric top angular momentum (p) replaced by the relative momentum of the top and frame $(p - \rho P)$. Thus, in the ρ axis coordinate system the Hamiltonian is similar in form to that for two coaxial symmetric tops, where the relative angular momentum is given by $(p - KI_\alpha/I_z)$. (Here K is the quantum number determining the component of total angular momentum along the figure axis and I_z is the moment of inertia about that axis.)

The coupling terms between the top and framework which appear to have been removed in this treatment, reappear in the guise of unusual boundary conditions on the Mathieu functions. The symmetry index of the torsional wavefunctions (σ), which has integer values for periodic functions, must now be written as $\sigma + KI_\alpha/I_z$, and in general is non-integral. That is, the appropriate Mathieu functions are no longer periodic in the angle α. However, solutions for these functions may be obtained from the periodic ones, and the energy levels may be expanded in a Fourier cosine series,

$$W_{v\sigma} = \sum_{n=0}^{\infty} w_n \cos \frac{2\pi n}{3} (\rho K - \sigma) \qquad (5.35)$$

The Fourier coefficients (w_n) depend only on the vibrational state (v) and the barrier parameter (s). For fairly high barriers $(s > 20$ for the ground torsional state) this series converges so rapidly that the terms with $n > 1$ become negligible. It is interesting to note that the convergence in this case is not dependent on ρ, as opposed to the corresponding power series in the PAM treatment. This implies that molecules with heavy tops (e.g. CF_3) rotating against light frameworks can be treated satisfactorily by the IAM approach.

Since the matrix elements of the internal rotation part of the Hamiltonian expressed in internal axes are much smaller than those of the rigid rotor matrix, the internal rotation energy may be considered as a perturbation on the rotational energy. The computation of a barrier height using the IAM is of necessity a computerized procedure and consists of matching the observed line splittings with those predicted for specific barrier values. The A and E predicted line frequencies are obtained by solving the torsional wave equation and adding the appropriate diagonal matrix elements to those of the rigid rotor Hamiltonian.

5.6 Comparison of Torsional Frequency and Splittings Methods

Microwave methods of studying molecules offer the advantage of always referring to isolated species in the vapour phase. The splittings method affords a further advantage in that all measurements are of frequencies, and these can be determined very accurately by microwave spectroscopy; to about 1 part in 10^6. Typically internal rotation splittings may be measured to ± 0.05 MHz when the components are separated by 2 100 MHz. Therefore any uncertainties in the values of potential barriers measured by this method have their origin in either, (a) the assumed molecular structure (i.e. the value of F) or, (b) the assumptions inherent in the molecular model (i.e. torsional mode being isolated from other vibrations) and not in the experimental data. The attainment of accurate intensity measurements in rotational spectroscopy is notoriously difficult, although the recent advent of commercial microwave spectrometers has significantly improved the situation. Even so, intensity measurements are seldom accurate to better than 10%. Frequency measurements in the infrared region (together with Raman spectroscopy) are also significantly less accurate than those in the microwave region; typical absorption bands being estimated to ± 1 cm^{-1} in 300 cm^{-1}. Another advantage of the microwave splittings method as compared with other methods is the fact that a much larger number of experimental measurements are available for use.

A comparison of the nature of the information derived from the splittings and torsional frequency methods is illustrated in Fig. 5.7. The latter method provides a value of $d^2V/d\alpha^2$ at V_{min} (i.e. the curvature of the potential energy well), whereas the former gives a measure of the area under the potential barrier.

Molecules with internal barriers in the range 2–8 kJ mol^{-1} are generally amenable to analysis by the splittings method. For this barrier range the splittings of the microwave lines in the ground vibrational state would vary from a few hundred MHz to a few hundredths of a MHz. Generally, internal rotation doublets falling in this range can be identified without undue difficulty. It is also possible to identify and measure internal rotation doublets in excited torsional states and so evaluate higher barriers by the splittings method, but in general such lines are considerably weaker than those of the ground vibrational state. It is interesting to note that for most molecules for which accurate microwave data is available a better fit between calculated and experimental spectral features is found if the methyl top is tilted by a few

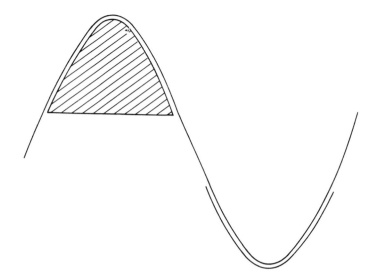

Fig. 5.7   Showing those parts of a potential function best determined by the splittings method (shaded area) and torsional frequency method (curvature of minimum).

degrees or so. Whether the methyl group tilts and retains its symmetric structure or whether there is asymmetry associated with methyl groups attached to asymmetric frameworks has not yet been ascertained. All barrier calculations are performed assuming a threefold symmetric top rotating about its figure axis.

Higher barriers are more easily measured by the torsional frequency method (or by NMR techniques) although there remains always the problem in vibrational spectroscopy of correctly assigning the absorptions to the appropriate torsional or vibrational state. Internal rotation barriers of symmetric top molecules cannot normally be investigated by the splittings method (see following section), but with certain exceptions (e.g. ethane) the torsional frequencies may be determined from their vibration spectra as described as Section 5.4.

5.7 Internal Rotation in Symmetric Rotors

For a rigid symmetric top molecule executing internal rotation the rotational lines are not affected by the internal motion since the energy levels of each J, K and $J + 1, K$ state are perturbed by exactly the same amount

(Fig. 5.4). Real molecules are, however, not rigid: this is illustrated clearly by the fact that rotational lines of excited vibrational states are found at frequencies shifted from those of the corresponding ground state lines. This satellite frequency pattern, based on non-rigidity effects, can be utilized to estimate the barrier to internal rotation for molecules. It is especially useful for symmetric top molecules where the number of determinable parameters is much smaller than for asymmetric molecules, and in addition symmetric top molecules cannot be analysed through the splittings method.

The observed pattern of rotational spectral lines is governed largely by the values of the three principal moments of inertia of a molecule. There is no direct relationship between the moments of inertia and the angle of internal rotation of a symmetric group, but for a non-rigid body the two are connected through the torsional motion interacting with other vibrations which in turn affect the rotational motion of a molecule.

The effective rotational constant of a symmetric rotor may be expressed in terms of a number of parameters which depend on the various intramolecular interactions that can occur. These parameters are evaluated as variables when the observed vibrational satellite frequencies are fitted to the effective rotational constant. They are then related directly to the height of the barrier hindering internal rotation. This method has been utilized successfully for certain symmetric top molecules[6] but for asymmetric molecules the corresponding analysis becomes very complex.

5.8 NMR Methods

The application of NMR spectroscopy to the determination of potential barriers is confined, by and large, to molecules with fairly high barriers ($V_3 > 25 \text{ kJ mol}^{-1}$). The technique represents unquestionably the most useful probe available for studying hindered internal rotation about bonds with partial double bond character (e.g. C—N link in amides). In principle, however, the NMR dynamic exchange method can be used for symmetric internal rotor groups such as CH_3. Provided that the spectral pattern of the internal rotor when rapidly rotating is different from that of the group when stationary, then NMR can give an upper (or lower) limit to the hindering potential (see Chapter 3).

For example, for monosubstituted ethanes (CH_3CH_2R), the three rotational isomers, corresponding to the thermodynamically most stable states at low temperatures (i.e. staggered) are all identical and show exactly the same spectral features. There exist three protons in different magnetic environments $(H_1 \equiv H_3 \neq H_2 \neq H_4 \equiv H_5)$ (Fig. 5.8).

Some of the coupling constants between the various protons are different, with the result that the overall NMR spectrum may be quite complicated —especially if the chemical shifts are of the same order of magnitude as the coupling constants. This would be true if the molecule were locked in the staggered conformation. If, on the other hand, the barrier is low and the internal rotation is fairly rapid then the magnetic screening and hence the chemical shifts of the three CH_3 protons will become the same, and there will exist one (average) coupling constant between the CH_3 and CH_2 group protons. Such a spectrum will be temperature independent.

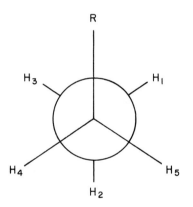

Fig. 5.8   The staggered conformation of a molecule $(CH_3 CH_2 R)$, showing the five protons, three of which, $H_1 (\equiv H_3)$, H_2 and $H_4 (\equiv H_5)$ are in different magnetic environments.

Under ambient conditions the protons of most methyl groups exchange so rapidly compared with the timescale of the NMR experiment that only one mean resonance signal is obtained. The movement may be slowed down in some cases by cooling the sample, but for methyl groups with barriers of the order of $12 \, kJ \, mol^{-1}$ or so, even cooling down to 170 K will not slow the process sufficiently for NMR spectroscopy to be useful.

Only for methyl groups in highly hindered environments can this method yield useful information about symmetric barriers and the method is much better suited for the study of asymmetric potentials and rotational isomerism as found, for example, in substituted amides.

5.9 Some Selected Methyl and other Symmetric Barriers

The heights of some potential barriers to internal rotation about single bonds, as measured by microwave spectroscopy and other techniques, are listed in Tables 5.2–5.8. These tables are not intended to be comprehensive, but rather, selected molecules have been arranged in tabular form so as to facilitate convenient comparisons of barriers within certain classes of compounds.[7] Much has been claimed, in the way of hypotheses relating to the origin of barriers, from the apparent trends in these barrier values. However, in general, as more experimental data becomes available so the pattern seems less clear cut (and more complex) and empirical deductions from such data become less reliable for the purpose of predicting other barriers.

Most of the tabulated data represent results obtained by microwave spectroscopy; this has the advantage that the values refer to vapour phase samples at fairly low pressures and so conditions approximate to those for isolated molecules. Barrier heights derived from torsional frequencies measured by far infrared or neutron diffraction studies on liquids or solids are generally found to be higher than the corresponding values for the vapour phase. This is believed to be caused by greater intermolecular interactions present in the condensed phases.

5.9.1 Barriers about C—C bonds in substituted ethanes

Monosubstituted ethanes retain the threefold symmetric nature of the potential energy function of ethane with respect to internal rotation, but the height of the barrier varies with substituent. For simple derivatives, the range of V_3 values (Table 5.2) is not large $(V_3 \text{(mean)} = 13{\cdot}8 \pm 1{\cdot}7 \text{ kJ mol}^{-1})$ even though the nature of the substituent differs considerably. Substituting a single hydrogen atom with any other simple group or atom generally produces a slightly higher barrier height than the value found for ethane. With more complex groups as substituents the methyl barrier is often considerably lower than $12{\cdot}5 \text{ kJ mol}^{-1}$. This is associated, in part, with the more flexible nature of the complex substituents and therefore greater vibrational interactions become likely between the C—C torsion and other molecular torsions and vibrations.

Substitution of a hydrogen atom of ethane by a halogen atom results in a higher barrier to internal rotation; this would appear logical in view of the larger size of the halogen atom. However, any hypothesis based on the suggestion that the potential barrier is a simple function of steric hindrance

Table 5.2   Potential barriers to internal rotation of some monosubstituted ethanes

Molecule	V_3 Barrier height (kJ mol^{-1})
$CH_3—CH_3$	12·25 (i.r.)[a]
$CH_3—CH_2F$	13·83
$CH_3—CH_2Cl$	15·42
$CH_3—CH_2Br$	15·41
$CH_3—CH_2I$	13·47
$CH_3—CH_2CN$	12·76
$CH_3—CH_2NC$	15·27
$CH_3—CH_2CH_2F$ (gauche)	11·99
$CH_3—CH_2CH_2F$ (trans)	11·25
$CH_3—CH_2CH_2Cl$ (gauche)	12·38 (i.r.)
$CH_3—CH_2CH_2Cl$ (trans)	11·63 (i.r.)
$CH_3—CH_2CHO$(cis)	9·54
$CH_3—CH_2COOH$	9·87
$CH_3—CH_2CH=CH_2$ (cis)	16·69
$CH_3—CH_2CH=CH_2$ (skew)	13·22
$CH_3—CH_2OH$	13·93
$CH_3—CH_2SH$	13·85 (t.d.)
$CH_3—CH_2CH_2Br$ (gauche)	11·09 (i.r.)
$CH_3—CH_2CH_2Br$ (trans)	9·87 (i.r.)
$CH_3—CH_2CH_2I$ (gauche)	11·59 (i.r.)
$CH_3—CH_2CH_2I$ (trans)	10·33 (i.r.)
$CH_3—CH_2C_6H_5$	11·30 (t.d.)
$CH_3—CH_2NH_2$	15·65
$CH_3—CH_2SiH_3$	11·09
$CH_3—CH_2GeH_3$	9·75
$CH_3—CH_2CHO$	9·54
$CH_3—CH_2CFO$	10·04

[a] All barriers measured by microwave spectroscopy except where indicated; i.r.—infrared, t.d.—thermodynamic.

is quickly shown to be untenable by the fact that addition of a second fluorine atom to the CH_2F group of $CH_3—CH_2F$ lowers, rather than raises, the barrier (Table 5.3). Nevertheless, the fact that steric factors can play an important role in determining barrier heights is demonstrated by the considerable difference observed between CH_3 barriers of different rotational isomers

Table 5.3   Potential barriers to internal rotation of some halogen substituted ethanes

Molecule	V_3 Barrier height (kJ mol^{-1})
$CH_3—CH_3$	12.25 (i.r.)[a]
$CH_3—CH_2F$	13·83
$CH_3—CHF_2$	13·31
$CH_3—CF_3$	13·60
CH_3 CH_2Cl	15·42
$CH_3—CHCl_2$	18·45
$CH_3—CCl_3$	20·92

[a] All barriers measured by microwave spectroscopy except where indicated; i.r.—infrared.

of the same molecule (e.g. *gauche-* and *trans*-1-fluoropropanes; Fig. 5.9; Table 5.2). The basic difference between these rotational isomers, which have the same molecular formulae, is the spatial orientation of the CH_3 group with respect to the rest of the molecule.

The V_3 values for the monohaloethanes (Table 5.2) show no obvious trend with size or electronegativity of the atom; $CH_3—CH_3 < I < F < Cl \simeq Br$. Factors other than the size of the substituents must be important in determining potential barriers as can be seen from the considerable difference in the barriers of the cyano CN and isocyano NC derivatives.

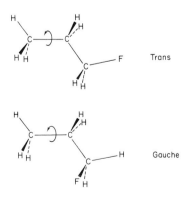

Fig. 5.9   *trans-* and *gauche*-1-fluoropropane.

Much work has obviously been done in attempting to correlate these experimental barrier values with concepts such as electronegativity, steric factors, etc., but there appears to be little doubt that the variation in V_3 values cannot be accounted for satisfactorily by invoking simple additivity principles. Indeed the apparent inconsistency of some of the trends have in themselves provided valuable data with which to test the validity of many other, more sophisticated, theories and hypotheses proposed to account for the origin of potential barriers (see, for example, Chapter 4).

5.9.2 Other symmetric barriers

When a methyl group is attached to an unsaturated carbon atom, as for example in propene or acetaldehyde, the potential barrier to internal rotation is lowered quite significantly from the value in ethane (Table 5.4), apparently reflecting the fact that for the unsaturated compounds the adjacent sp^2 hybrid bonds are pushed further away from the methyl group. As for monosubstituted ethanes, the potential barriers of the methyl groups for a series of substituted aldehydes are remarkably similar with the notable exceptions of acetic acid and methyl acetate. The low barriers associated with these two compounds is almost certainly due to the increased symmetry of the frameworks about which the methyl groups rotate,

and in the extreme case of the ion the barrier would become sixfold,

When rotation about single bonds other than C—C bonds is considered the range of values found for methyl top barriers is considerably extended. Table 5.5 shows the effect of replacing one of the CH_3 groups in ethane by the analogous groups SiH_3, GeH_3 and SnH_3. The threefold potential barrier becomes progressively lower with the heavier groups. A similar trend is found to occur within the series NH_2, PH_2 and AsH_2 when these groups are attached to a CH_3 group.

Table 5.4   Potential barriers to internal rotation of some CX_3 groups adjacent to a carbonyl bond

Molecule	V_3 Barrier height (kJ mol^{-1})
$CH_3—C{\Large\langle}^O_H$	4·89[a]
$CH_3—C{\Large\langle}^O_F$	4·36
$CH_3—C{\Large\langle}^O_{CH_3}$	3·26
$CH_3—C{\Large\langle}^O_{Cl}$	5·42
$CH_3—C{\Large\langle}^O_{Br}$	5·46
$CH_3—C{\Large\langle}^O_I$	5·44
$CH_3—C{\Large\langle}^O_{CN}$	5·06
$CH_3—C{\Large\langle}^O_{OH}$	2·02
$CH_3—C{\Large\langle}^O_{OCH_3}$	1·25
$CH_3—C{\Large\langle}^O_{CCH}$	4·49
$CF_3—C{\Large\langle}^O_H$	3·70
$CH_3—C{\Large\langle}^O_{CHCH_2}$	5·23

[a] All barriers measured by microwave spectroscopy.

However, even this trend of decreasing methyl barrier as a particular group of the Periodic Table is descended does not hold for all groups. The OH, SH and SeH series shows anomalous behaviour in this respect, clearly

Table 5.5   Potential barriers to internal rotation of CH_3 groups attached to various elements within the same group of the Periodic Table

Molecule	V_3 Barrier height (kJ mol^{-1})
CH_3-CH_3	12·25 (i.r.)[a]
CH_3-SiH_3	6·97
CH_3-GeH_3	5·18
CH_3-SnH_3	2·72
CH_3-NH_2	8·27
CH_3-PH_2	8·20
CH_3-AsH_2	6·19
CH_3-AsF_2	5·54
CH_3-OH	4·48
CH_3-SH	5·31
CH_3-SeH	4·23

[a] All barriers measured by microwave spectroscopy except where indicated; i.r.—infrared.

indicating that increased bond length and atomic size are not the only factors affecting the barrier heights. A comparison of methyl barriers about C—O and C—S bonds (Table 5.6) also emphasizes the lack of any obvious trend in barriers for corresponding oxygen and sulphur compounds. The involvement of "d" orbitals in sulphur has been proposed for some of the anomalous values for many sulphur derivatives.

Table 5.6   Comparison of the potential barriers hindering internal rotation of CH_3 groups attached to oxygen and sulphur

Molecule	V_3 Barrier height (kJ mol^{-1})	
	X = O	X = S
$CH_3-X(CH_3)$	11·38[a]	8·92
$CH_3-X(CCH)$	6·03	7·32
$CH_3-X(Cl)$	12·80	10·88
$CH_3-X(H)$	4·48	5·31
$CH_3-X(CN)$	4·64	6·65
$CH_3-X(CHO)$	4·98	>8·4
$CH_3-X(CHCH_2)$	16·02	13·51

[a] All barriers measured by microwave spectroscopy.

5.10 Low Potential Barriers

Most methyl group barriers have been calculated using the assumption that in the expansion of the hindering potential $V(\alpha)$ all terms higher than V_3 may be neglected or assumed zero. This would appear to be a reasonable assumption in view of the large number of molecules whose spectral splittings can be accurately predicted from the calculated V_3 barrier heights. The next higher term in the series becomes particularly noticeable only when V_3 is zero, and experimental values for V_6 have been obtained for certain molecules where symmetry precludes a threefold potential (e.g. toluene, nitromethane). Experimental data confirm the fact that V_6 barriers are very much lower than V_3 (Table 5.7).

The occurrence of a sixfold barrier for the methyl group in nitromethane may be seen from Fig. 5.10. It can be seen that the presence of an additional V_6 term will not affect the barrier height of a molecule with a dominant threefold potential since the V_6 term has minima at both maxima and minima points on the V_3 curve. However, a V_6 term can affect the overall shape of the dominant potential function depending on the relative phases of the V_3 and V_6 terms. In Fig. 5.10, both functions are in phase and the effect of adding

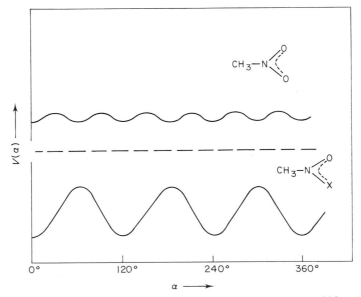

Fig. 5.10   The effect of internal rotation of a CH₃ group against NO₂ and NOX frameworks. The threefold effect on the potential function is missing in the former.

V_3 and V_6 terms results in narrower potential wells. This is further illustrated in Fig. 5.11, where it is also seen that a negative V_6 contribution (i.e. out of phase with the V_3 function) would broaden the energy wells. Consequently the torsional energy levels within the potential wells are pushed further apart by a positive V_6 and more closely together by a negative V_6 contribution.

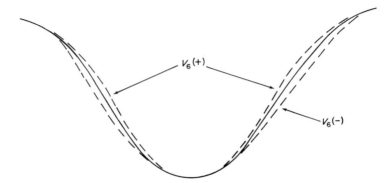

Fig. 5.11   Showing the effect of a positive and negative V_6 term on the shape of part of a threefold potential function.

The experimental determination of meaningful values for the leading term of the Fourier expansion in $V(\alpha)$, whether it be V_3 or V_6, is relatively straight-forward, but there is much more uncertainty associated with the acquisition of both V_3 and V_6 terms for the same molecule. To evaluate the first two terms of the Fourier expansion requires data gathered from excited torsional states in addition to the ground state. Since the V_6 contribution to the barrier is generally much smaller than the V_3 effect, it can be treated as a perturbation. When data from excited torsional states are used, the value of V_3 alone, as determined by the procedure outlined previously in this chapter, is often found to vary for different torsional states. Incorporation of a finite V_6 term, however, enables a constant V_3 value to be derived. One convenient procedure of introducing a V_6 term consists of utilizing exactly the same equations, etc., as for the determination of a "pure" V_3 term, but to use modi-fied values of the perturbation coefficients $W_{v\sigma}$. The eigenvalues of Mathieu's equation ($b_{v\sigma}$) may be expressed as follows,[8]

$$b'_{v\sigma} = b_{v\sigma} + \left(\frac{V_6}{V_3}\right)\frac{s}{2}(1 - \langle v|\cos 6\alpha|v\rangle) + \left(\frac{V_6}{V_3}\right)^2\frac{s^2}{4}\left(\frac{66}{\Delta}\right) \qquad (5.36)$$

where $b'_{v\sigma}$ is the modified eigenvalue. The term $(66/\Delta)$ represents the second order perturbation contribution

$$\left(\frac{66}{\Delta} = \sum_{v'} \frac{\langle v'|\cos 6\alpha|v\rangle \langle v|\cos 6\alpha|v'\rangle}{b_v - b_{v'}}\right),$$

and both first ($\langle v|\cos 6\alpha|v\rangle$) and second order correction terms are readily available in tabulated form. This expression is accurate for V_6/V_3 of the order of 0·05 or smaller. The correct value of V_6/V_3 is taken as that which gives a constant value of V_3 for all vibrational states.

Table 5.7   Some molecules with low V_3 and V_6 potential barriers

Molecule	V_6 Barrier height (J mol^{-1})
CH_3—⬡	58·32[a]
CH_3—⬡—F	57·82
CH_3—⬡—Cl	58·16
CH_3—⬡N	56·48
CH_3—NO_2	25·23
CD_3—NO_2	21·71
CF_3—NO_2	311·3
CH_3—BF_2	57·61

Molecule	V_3 (kJ mol^{-1})	V_6/V_3 (%)
CH_3—CH_2Cl	15·44	0
CH_3—$CHCH_2$	$\begin{cases} 8·35 \\ 8·49 \end{cases}$	$\begin{cases} -1·9 \\ -2·3 \text{ (i.r.)} \end{cases}$
CH_3—$CFCH_2$	10·21	0·6
CH_3—CH—CH_2 (O)	10·71	< 2·5
CH_3—⬡—F	either $\begin{cases} 0·19^b \\ 0·20 \end{cases}$ or	$\begin{cases} 26·5 \\ -15·4 \end{cases}$

[a] All barriers measured by microwave spectroscopy except where indicated; i.r.—infrared.
[b] m fluoroanisole represents an example where V_3 and V_6 are of comparable magnitude[13].

For most examples where it is claimed that both V_3 and V_6 terms have been evaluated, the ratio V_6/V_3 is lower than 3% (Table 5.7), It should be noted, however, that a variation in moment of inertia of the top with torsional angle can produce a similar effect to the inclusion of a V_6 term in the potential function. Such an effect might occur if the internal rotation were coupled to another low lying vibration. Indeed, the whole question of the physical significance of the V_6 terms reported in the literature when measured in the presence of V_3 terms is uncertain, owing to the apparent equivalence of the V_6 term to torsional flexing which may occur in molecules with internal rotation.[9] The effect of such torsional flexing is likely to be greatest for molecules with "light" frameworks (e.g. methanol, methylamine, etc.).

One feature of possible significance with regard to this uncertainty is the fact that for those molecules with sixfold potential symmetry, finite V_6 values have been measured and these are of the same order of magnitude as those found when V_6 is determined in the presence of a V_3 term.

5.11 Multiple Symmetric Groups

The expressions relating the potential barrier to the angle of internal rotation for molecules with more than one symmetric group are more complex than those for the single top case.[10] The two rotors may interact with each other as well as with the rest of the molecular framework. In the extreme case of a molecule having two methyl groups in quite different environments and with very different barrier heights (e.g. CH_3COOCH_3 where the potential barriers of the two methyl groups are $1 \cdot 28$ and $5 \cdot 08 \, kJ \, mol^{-1}$) there is evidence to suggest that the rotors may be considered quite independently of each other, and the problem reduces to that of two independent methyl groups, each affecting the rotational spectrum as if the other were not present. In many instances, however, the methyl groups are equivalent and must be treated together (e.g. $(CH_3)_2O$ and $(CH_3)_2CO$).

The Hamiltonian for the overall and internal rotation in such cases may be considered to consist of the following components,

$$H = H_R + H_T + H_{TT} + H_{RT}$$

H_R represents the pure rotational Hamiltonian, H_T the internal rotation, H_{TT} the interaction between the tops, and H_{RT} the interaction between the internal and overall rotation. For systems containing two equivalent methyl groups the respective terms of the Hamiltonian have the form,

$$H_R = AP_a^2 + BP_c^2 + CP_c^2 + F(\mathscr{P}_1^2 + \mathscr{P}_2^2) + F'(\mathscr{P}_1\mathscr{P}_2 + \mathscr{P}_2\mathscr{P}_1)$$

$$H_T = F(p_1^2 + p_2^2) + \frac{V_3}{2}(1 - \cos 3\alpha_1) + \frac{V_3}{2}(1 - \cos 3\alpha_2)$$

$$H_{TT} = F'(p_1p_2 + p_2p_1) + V_3' \cos 3\alpha_1 \cos 3\alpha_2 + V_3'' \sin 3\alpha_1 \sin 3\alpha_2$$
$$+ \frac{V^*}{2}(\cos 6\alpha_1 + \cos 6\alpha_2) + \dots$$

$$H_{RT} = -2F(p_1\mathscr{P}_1 + p_2\mathscr{P}_2) - 2F'(p_1\mathscr{P}_2 + p_2\mathscr{P}_1) \qquad (5.37)$$

The symbols α_1 and α_2 refer to the torsional angles of the two tops, while $\mathscr{P}_1$ and $\mathscr{P}_2$ are defined as,

$$\mathscr{P}_1 = I_\alpha\left(\frac{\lambda_a P_a}{I_a} + \frac{\lambda_b P_b}{I_b}\right)$$

$$\mathscr{P}_2 = I_\alpha\left(\frac{\lambda_b P_b}{I_b} - \frac{\lambda_a P_a}{I_a}\right) \qquad (5.38)$$

F and F' are the reduced rotational constants, and take the form,

$$F = \frac{\hbar}{4I_\alpha}\left[\frac{I_a}{(I_a - 2\lambda_a^2 I_\alpha)} + \frac{I_b}{(I_b - 2\lambda_b^2 I_\alpha)}\right)$$

and

$$F' = \frac{\hbar}{4I_\alpha}\left[\frac{I_b}{(I_b - 2\lambda_b^2 I_\alpha)} - \frac{I_a}{(I_a - 2\lambda_a^2 I_\alpha)}\right]$$

respectively. V_3, V^*, V_3' and V_3'' represent coefficients of the Fourier expansion of the potential energy up to second order.

The interaction terms H_{TT} and H_{RT} are considerably smaller than the pure torsion and rotational energy terms, and for analysis of the torsional modes in the far infrared spectral region, the assumption is also made that the effect of the H_{RT} term is negligible. This latter term, however, is important in microwave studies, since this coupling between the torsion and overall rotation gives rise to measurable effects on the pure rotation spectra. In microwave studies of the ground vibrational state of double top molecules,

H_{TT} is often assumed to be zero, or so small that it can be neglected (i.e. $F' = V_3' = V_3'' = V^* = 0$). However, small but finite top-top interactions can be treated as perturbations using a PAM approach analogous to the perturbation treatment for single top molecules.

Detailed consideration of the symmetry properties of the Hamiltonian shows that the torsional energy levels in the ground vibrational state of a two-top molecule are ninefold degenerate in the limit when both barriers are very high, but as the barriers become lower, this degeneracy is partially lifted to give four levels which are labelled AA, EE, AE and EA. The AA level

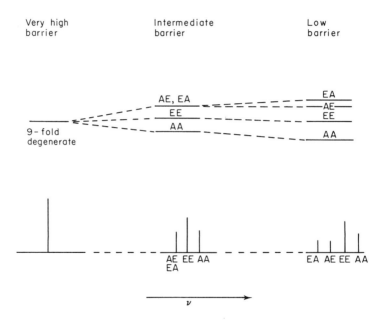

Fig. 5.12   Showing the effect of internal rotation on the degeneracies of the torsional states of molecules with two methyl rotors, and the approximate splitting pattern of the rotational transitions.

is non-degenerate, the AE and EA states are each doubly degenerate, while the EE state is quadruply degenerate. Symmetry selection rules may be derived for spectra of molecules with double tops in a manner similar to that outlined in Section 5.3.1 for single rotors, though the group theoretical arguments are more complex owing to the higher symmetry. The predicted quartet absorption pattern is observed for many rotational transitions of acetone for example where $V_3 = 3.26\ \text{kJ mol}^{-1}$, whereas for molecules with somewhat higher barriers, e.g. dimethyl ether, where $V_3 = 11\cdot38\ \text{kJ mol}^{-1}$ the ground state rotational spectrum consists of many triplet absorption lines. For these intermediate values of the barrier (i.e. $V_3 \sim 10\ \text{kJ mol}^{-1}$) the two levels AE and EA remain degenerate. The separation between the components of the observed triplets or quartets are sensitive functions of the barrier height.

Table 5.8   Potential barriers to internal rotation of molecules with multiple CH_3 groups

Molecule	Barrier height (kJ mol^{-1})
CH_3 CH_3 >CH_2	$13\cdot9$[a]
CH_3 CH_3 >CHF	$14\cdot9$
CH_3 CH_3 >$CHCl$	$14\cdot7$
CH_3 CH_3 >$CHBr$	$17\cdot8$ (i.r.)
CH_3 CH_3 >CHI	$17\cdot3$ (i.r.)
CH_3 CH_3 >$CHCN$	$13\cdot8$
CH_3 CH_3 >$CHOH$	$14\cdot2$ (t.d.)
CH_3 CH_3 >$C{=}O$	$3\cdot26$
CH_3 CH_3 >O	$11\cdot38$
$(CH_3)_3CH$	$16\cdot3$
$(CH_3)_3CCCH$	$17\cdot2$ (i.r.)
$(CH_3)_3CCN$	$18\cdot0$ (i.r.)
$(CH_3)_3CCl$	$18\cdot9$

[a] All barriers measured by microwave spectroscopy except where indicated; i.r.—infrared, t.d.—thermodynamic.

Vibrational studies of multiple rotor molecules can be used to derive potential barriers from direct observation of the torsional modes in the far-infrared or Raman spectra. A double top molecule has two torsional modes and for a molecule with a planar framework or an asymmetric framework both torsions are active in the infrared and Raman. For more symmetric molecules, e.g. those belonging to the point group C_{2v}, one of the torsional modes becomes inactive in the infrared. However, clear-cut deductions regarding potential barriers from torsional frequencies are often complicated by the presence of "hot" bands originating from excited torsional states (e.g. $v = 2 \leftarrow v = 1$). These transitions generally fall fairly close to the fundamental $v = 1 \leftarrow v = 0$ modes and for low frequency torsions their intensities can be comparable to that of the fundamental.

Table 5.8 lists values for the potential barriers for some molecules of the substituted propane series. Although slightly higher than the values for single top molecules, it is interesting that the range is close to that of the latter series.

When three methyl groups are attached to the same carbon atom, steric forces are larger, resulting in higher potential barriers to internal rotation, Table 5.8. The barriers tend to be too high for internal rotation splittings to be detected in the microwave spectra of either the ground state or the first excited torsional states (e.g. $(CH_3)_3CCHO$), and so most of the data relating to barriers of methyl groups in three-top molecules has some from analysis of the torsional modes through far-infrared and Raman spectroscopy. The intertop coupling in such molecules is generally estimated to be about 10% of the main V_3 value.

5.12 Internal Rotation in Methanol and Acetaldehyde

In this final section we shall briefly consider internal rotation in two molecules, methanol and acetaldehyde, to show the effect on the potential barrier of:

(a) interaction between the torsional mode and other low frequency vibrations, and

(b) slight asymmetry of the internal rotor.

These molecules have been selected because of the unique features associated with their internal rotation and as such they do not represent truly typical examples of molecules exhibiting internal rotation.

5.12.1 Methyl alcohol

Methanol represents a molecule which has a symmetric rotor (CH_3) attached to a planar framework (OH). The shape of the potential energy curve describing the hindered internal rotation is similar to that of any threefold symmetric rotor (Fig. 5.2). However, the situation in methanol is made more complicated by the fact that the framework is lighter than the symmetric rotor (i.e. in methanol it is a case of the "tail wagging the dog"!). The rotational spectrum of the molecule has been the subject of several very detailed analyses,[11] and it is was realized at an early date that owing to the heavy top and light framework, the best theoretical approach to the problem lay in the IAM treatment. (The currently accepted value for the barrier in methanol is $4.47 \, kJ \, mol^{-1}$.).

It has not been found possible to fit all of the observed microwave splittings accurately using the usual first-order barrier treatment. More refined calculations have included the effects of interactions between the torsional mode and other molecular vibrations. Higher terms in the potential energy function have been proposed for methanol ($V_6 = 0 - 10 \, J \, mol^{-1}$), but it has subsequently been shown that torsional flexing of the molecule during internal rotation contributes terms of this order of magnitude.

The barrier height has also been determined for excited vibrational states of methanol. The value of V_3 for the first excited state of the "CH_3 rocking" mode which lies at $933 \, cm^{-1}$ above the ground state, is $6.66 \pm 0.42 \, kJ \, mol^{-1}$. This represents a dramatic increase of 50% over the value obtained for the ground state. A smaller increase in the value of V_3 has been found for the first excited state of the C—O stretching mode, where V_3 is $4.76 \pm 0.04 \, kJ \, mol^{-1}$. These observed changes in V_3 for methanol in going from the ground state to excited vibrational states are too large to be attributable to experimental errors, and so it might be concluded that the V_3 value derived from the ground state spectrum probably contains a sizeable contribution from zero-point vibration effects. This implies that comparison of experimental barriers for molecules of this type with those derived from *ab initio* calculations may not be too meaningful since the latter are often computed for a rigid model with equilibrium parameters.

5.12.2 Acetaldehyde

Acetaldehyde has a symmetric threefold potential and the microwave spectrum of the normal species can be fitted very well with inclusion of only the leading V_3 term of the potential energy series. However, much interesting

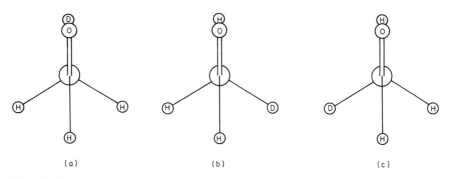

(a)   (b)   (c)

Fig. 5.13   The three stable conformations of monodeutero-acetaldehyde, showing the two equivalent *gauche* structures (b) and (c).

work has been carried out on the various deuterated species of acetaldehyde.[12] For the molecule CH_2DCHO (or CHD_2CHO), the methyl group is no longer threefold symmetric, and Fig. 5.13 shows that there can exist three separate rotational isomers.

The two *gauche* conformations (b) and (c) are identical from an energetic point of view. Owing to the similar electronic properties of C—H and C—D bonds the potential function for internal rotation in this molecule may be assumed to retain its threefold symmetry (Fig. 5.14). However, the torsional energy levels within the potential energy wells are affected by the change in mass on partial deuteration, and the molecular kinetic energy varies as a function of the angle of rotation.

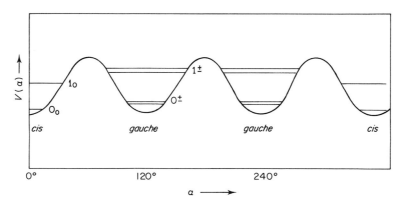

Fig. 5.14   Potential function and torsional energy levels for deutero-acetaldehyde.

Each torsional state, which in normal acetaldehyde is split into an A and E doublet, is now split into three components, and each microwave absorption line is a triplet compared with the usual doublet. One possible interpretation of this is that one of the rotational lines arises from the rotor with C_s symmetry (Fig. 5.13a), whilst the other two lines are caused by tunnelling through the barrier separating the two equivalent *gauche* rotamers (Figs 5.13b and 5.13c).

A fuller account of how asymmetry in the rotor affects the internal motions of molecules including molecules which exist as quite distinct rotational isomers is given in the following chapter.

Further Reading

1. C. C. Lin and J. D. Swalen. Internal Rotation and Microwave Spectroscopy, *Reviews of Modern Physics*, **31**, 841 (1959).
   One of the first good comprehensive reviews on internal rotation in molecules.
2. W. Gordy and R. L. Cook. "Microwave Molecular Spectra". Wiley, Chichester and New York, 1970.
   This is an excellent modern text dealing with microwave spectroscopy in which internal rotation of symmetric groups is treated in some detail.
3. J. E. Wollrab. "Rotational Spectra and Molecular Structure". Academic Press, London and New York, 1967.
   Another detailed treatment of internal rotation of symmetric groups including the internal axis method (IAM).

References

1. L. G. Smith. *J. Chem. Phys.* **17**, 139 (1949).
2. S. Weiss and G. E. Leroi. *J. Chem. Phys.* **48**, 962 (1968).
3. (a) E. Hirota and C. Matsumura. *J. Chem. Phys.* **55**, 981 (1971).
   (b) E. Hirota, K. Matsumura, M. Imachi, M. Fujio, Y. Tsuno and C. Matsumura. *J. Chem. Phys.* **66**, 2660 (1977).
4. (a) H. C. Longuet-Higgins. *Mol. Phys.* **6**, 445 (1963).
   (b) J. K. Watson. *Canad. J. Phys.* **43**, 1996 (1965).
5. R. C. Woods III. *J. Mol. Spectroscopy*, **21**, 4 (1966); **22**, 49 (1967).
6. V. W. Laurie. *J. Mol. Spectroscopy*, **13**, 283 (1964).
7. For more comprehensive tables of barriers the reader is directed to:
   (a) W. Gordy and R. L. Cook. "Microwave Molecular Spectra", Chap. 12. Wiley, Chichester and New York, 1970.
   (b) J. E. Wollrab. "Rotational Spectra and Molecular Structure", Appendix 9. Academic Press, London and New York, 1967.
   (c) J. R. Durig, S. M. Craven and W. C. Harris. Determination of torsional barriers from far-infrared spectra, p. 73. *in* "Vibrational Spectra and Structure" (J. R. Durig. Ed.), Vol. 1, Dekker, New York, 1972.

(d) J. P. Lowe, Barriers to internal rotation *in* "Progress in Physical Organic Chemistry" (A. Streitwisser Jr. and R. W. Taft, Eds.), Vol. 6. Wiley, Chichester and New York, 1968.

8. D. R. Herschbach. "Tables for the Internal Rotation Problem". Dept. of Chemistry, Harvard University, 1957.

9. (a) C. S. Ewing and D. O. Harris. *J. Chem. Phys.* **52**, 6268 (1970).
    (b) R. M. Lees. *J. Chem. Phys.* **59**, 2690 (1973).

10. J. E. Wollrab. "Rotational Spectra and Molecular Structure", p. 194. Academic Press, London and New York, 1967.

11. R. M. Lees and J. G. Baker. *J. Chem. Phys.* **48**, 5299 (1968).

12. C. R. Quade and C. C. Lin. *J. Chem. Phys.* **38**, 540 (1963).

13. H. D. Rudolph and A. Trinkaus. *Z. Naturforschg.* **23a**, 68 (1968).

6
Internal Rotation of Asymmetric Groups

6.1 Introduction

From the discussion in the previous chapter, it is evident that much information, often highly detailed, has been built up about internal rotation in molecules in which the internal rotor has the rotational properties of a symmetric top. However, in many molecules in which internal rotation is possible, the internal top does not have the threefold or higher symmetry of a symmetric top and much less information is currently available about the details of internal rotation in these cases. The reason for this probably lies largely in the fact that the dynamics of internal rotation involving an asymmetric group is considerably more difficult to handle than for a symmetric internal rotor. In this chapter we examine the main consequences of a reduction in the symmetry of the internal top.

Molecules in which internal rotation can occur around a bond joining two groups, neither of which is a symmetric rotor, have the possibility of existing in different conformations. This has already been mentioned in Chapter 5 in connection with partially deuterated acetaldehyde. In this case, since only isotopic substitution is involved, the potential function for the partially deuterated molecule retains the threefold symmetry of that for acetaldehyde itself. Often different rotamers of a molecule have different relative stabilities, and this is reflected in the shape of the potential function governing internal rotation. In propionyl fluoride, for example, where rotation around the C—CFO bond can occur, the rotamers labelled *cis* and *gauche* (Fig. 6.1) have been shown to coexist, the *cis* being more stable. The shape of the experimentally determined potential function is also shown in Fig. 6.1 and clearly no longer has the threefold symmetry of the corresponding curve for internal rotation of a methyl group. As described in Chapter 2 the usual method of

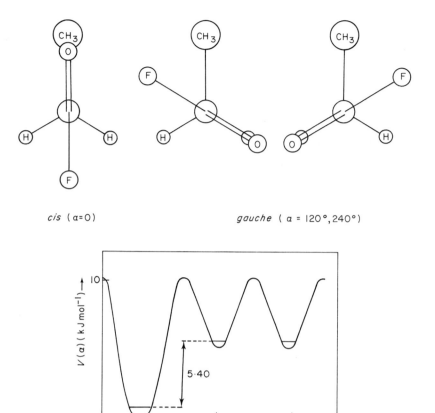

Fig. 6.1   The stable rotamers and experimental potential function for propionyl fluoride.

describing curves of this type is to express the potential energy (V) as a Fourier series in the internal rotation angle (α) (equation (2.42)). The larger number of parameters required to define the shape of the curve for an asymmetric internal rotor presents an apparent drawback to the study of internal rotation in these more complex cases. Fortunately, however, this is offset to some extent by the fact that more barrier-dependent data are often experimentally determinable. For example, energy differences between different rotameric forms of a molecule may be determined spectroscopically or, in favourable cases, by gas phase electron diffraction. Torsional vibrational frequencies and sub-state splittings can often be determined from infrared, far-infrared or microwave spectra, or rotational constants may be found to vary with the torsional state in a manner which can be related to the shape of the potential function.

In addition to introducing the possibility of the existence of different, stable, molecular conformations, there is a further important difference between internal rotation of an asymmetric group and its symmetric counterpart. As mentioned in Chapter 5, for the internal rotation of a symmetric top group the moments of inertia of the entire molecule are independent of the relative orientation of the top and frame. However, when an asymmetric internal top rotates the moments of inertia become strongly dependent on the orientation of the top, the same being true for the reduced moment of inertia for the torsional vibration. This angular dependence of the inertial parameters of a molecule makes the computation of energy levels, wave functions and potential functions a rather more involved process than in the case of a symmetric internal top.

The combined effect of the additional terms necessary to define the potential energy function for an asymmetric internal rotor and the torsional dependence of the inertial parameters of the molecule can lead to considerable deviation of the pattern of torsional energy levels from that described in Chapter 5 for the symmetric internal top. This is illustrated qualitatively in Fig. 6.2. For simplicity the addition of only one term, V_2, to the threefold symmetric potential function is considered.

The effect on the shape of the potential function of the addition of the V_2 term to the V_3 term in the same phase is shown in Fig. 6.2(a). The extent of the deviation of the curve from that of Fig. 5.1 depends on the relative magnitudes of V_2 and V_3. It can be seen from Fig. 6.2(c) that the effect on the energy levels of the addition of the twofold term is to generate a set of singly degenerate levels of alternating parity and a set of nearly degenerate pairs of levels each member of a pair with opposite parity. The single levels are associated with the potential well at $\alpha = 0°$ and the nearly degenerate pairs with the identical minima on either side of $\alpha = 180°$. The properties of the energy levels in the identical minima are qualitatively similar to those of the energy levels in the double minimum potential functions associated with inversion (Chapter 7).

The effect produced by introducing the angular dependence of the reduced rotational constant is shown in Fig. 6.2(d). In particular it will be noticed that the energy separations between the pairs of levels in the two identical minima are reduced whilst those between the single levels are increased. Also it is interesting to note that the degenerate E levels of the pure threefold symmetric potential function (Fig. 6.2(a)) do not necessarily correspond to the nearly degenerate pairs of Fig. 6.2(d).

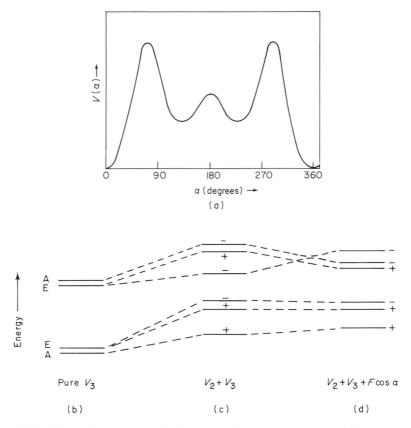

Fig. 6.2   Effect of asymmetry in the internal rotor on the potential function and energy levels of internal rotation. (a) Illustrates the shape of the function obtained by adding a $V_2 \cos 2\alpha$ term to a $V_3 \cos 3\alpha$ function in phase; (b) shows the torsional manifold of a pure threefold potential; (c) shows the torsional manifold of a potential function of the type shown in (a); (d) shows the effect of a variable reduced mass on the torsional levels.

6.2 Asymmetric Potential Functions

The mathematical formulation of periodic potential functions has already been discussed in Chapter 2. In order to determine a sufficient number of coefficients (V_n) in the expansion for the potential energy of an asymmetric internal rotor (equation (2.42)) it is often necessary to combine data from several different experimental sources. When this is done the varying accuracy with which the barrier dependent data has been obtained by the different experimental techniques must be carefully noted. In practice, the details of the determination of the numerical coefficients of equation (2.42) for a particular molecule vary from case to case depending on the form of the experimental data.

A number of simple relationships between the coefficients (V_n) which are useful in the determination of asymmetric potential functions can be deduced from the expression for $V(\alpha)$. A stable rotamer of a molecule is associated with a minimum in the potential function for internal rotation. Consequently, when $\alpha = \alpha_e$, where α_e is the equilibrium internal rotation angle for the rotamer, the first derivative of $V(\alpha)$ must vanish,

$$\left(\frac{\partial V(\alpha)}{\partial \alpha}\right)_{\alpha_e} = \sum_n n\frac{V_n}{2}\sin n\alpha_e = 0 \tag{6.1}$$

A second relationship between the coefficients is obtained by using the equivalence between the second derivative of $V(\alpha)$ and the harmonic force constant,

$$\left(\frac{\partial^2 V(\alpha)}{\partial \alpha^2}\right)_{\alpha_e} = \sum_n \frac{n^2 V_n}{2}\cos n\alpha_e = \left(\frac{v^2}{2F}\right)_{\alpha_e} \tag{6.2}$$

where v is the appropriate fundamental vibration frequency and F is the corresponding reduced rotational constant for the vibration. Often, the available experimental data includes energy differences between ground vibrational states of the stable rotamers. The energy difference (ΔE) between any two rotamers A and B, after correcting for zero point vibrational effects, may be equated to the difference between the values of $V(\alpha)$ at the equilibrium configurations $\alpha_e(A)$ and $\alpha_e(B)$,

$$\sum_n \frac{V_n}{2}(\cos n\alpha_e(A) - \cos n\alpha_e(B)) = \Delta E + \tfrac{1}{2}(hv(A) - hv(B)) \tag{6.3}$$

Additional information about the shape of the potential function can be obtained from an investigation of the anharmonicity of the potential well associated with a particular rotamer provided a sufficient number of vibrational states of the rotamer have been studied to warrant this. The variation of rotational constants with torsional state can be useful in this respect.

With the exception of the angles α_e, the experimental data used in equations (6.1–6.3) can be obtained by relative intensity measurements on spectra of the compound under investigation. As mentioned in Chapter 3, spectroscopic relative intensity measurements are often subject to large experimental errors

F

and these will be transmitted to potential constants derived using this procedure. In addition to the sensitivity to experimental error another factor which contributes to the uncertainty of the derived potential function is the fact that the experimental data in many cases refers only to those parts of the potential function close to the bottom of the potential wells associated with stable rotamers. Data of this kind cannot yield definitive information about the potential curve in the regions around the tops of the barriers. This difficulty can be overcome by obtaining data related to vibrational states which lie close to the tops of the barriers. Unfortunately, since these energy levels are often of relatively high energies they are thinly populated and the associated spectra may well be too weak to be identified with any certainty.

A fairly typical illustration of the determination of a potential function by the method outlined above is provided by 1,1,2,2-tetrabromoethane.[1] Data obtained from the far-infrared and Raman spectra of the stable *gauche* and *trans* forms of this molecule (Fig. 6.3) pertaining to the shape of the potential function for internal rotation are summarized in Table 6.1. Direct substitution into equations (6.1–6.3), with the *trans* rotamer assigned an internal

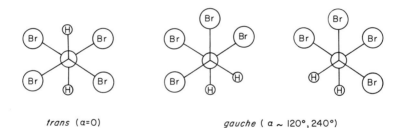

trans ($a=0$)   *gauche* ($a \sim 120°, 240°$)

Fig. 6.3   *Trans* and *gauche* rotamers of 1,1,2,2-tetrabromoethane

rotation equilibrium angle of zero, yields four equations which allow the determination of the first three coefficients V_n of the potential function and the equilibrium angle of the *gauche* rotamer. It can be seen from Table 6.1 that the resulting potential constants are dominated by the V_3 term which is not surprising since this is the only one of the first three terms which has minima at $\alpha = 0°$ and $\alpha = 120°$.

Another example of a potential function derived in a similar manner is that for internal rotation around the $C—CO_2H$ bond of acrylic acid.[2] The stable forms of this molecule are the s-*cis* and s-*trans* rotamers (Fig. 6.4) the former being slightly the more stable. The experimental data in this case are derived solely from the microwave spectrum of the molecule and are given in Table 6.2.

Table 6.1   Experimental data relating to the potential function for internal rotation in 1,1,2,2-tetrabromoethane

α_e (trans)	0°
F (trans)[a]	0·0812 cm^{-1}
F (gauche) ($\alpha_e = 113°$)[a]	0·0742 cm^{-1}
v (trans)	35 cm^{-1}
v (gauche)	38 cm^{-1}
ΔE (trans–gauche)	2·85 kJ mol^{-1} (235 cm^{-1})
V_1	9·1 kJ mol^{-1}
V_2	−11·8 kJ mol^{-1}
V_3	24·2 kJ mol^{-1}
α_e (gauche)	113°

[a] Calculated from an assumed model.

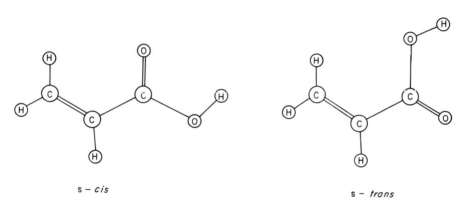

s – cis

s – trans

Fig. 6.4   The stable conformations of acrylic acid.

Table 6.2   Experimental data pertaining to the shape of the potential function for internal rotation in acrylic acid

α_e(s-cis)	0°
α_e (s-trans)	180°
F (s-cis)[a]	1·917 cm^{-1}
F (s-trans)[a]	1·900 cm^{-1}
v (s- cis)	105 ± 20 cm^{-1}
v (s-trans)	95 ± 20 cm^{-1}
ΔE (s-trans–s-cis)	0·69 ± 0·24 kJ mol^{-1} (58 ± 20 cm^{-1})

[a] Calculated from an assumed structure.

Only three pieces of experimental data are available so the Fourier series for the potential energy is truncated at the term V_3 and use of equations (6.2) and (6.3) with the s-*cis* rotamer assigned $\alpha_e = 0$ and the s-*trans* $\alpha_e = 180°$ enables the three potential constants V_1 to V_3 to be determined. The resulting potential function has the form shown in Fig. 6.5 and it will be noticed that, as expected, the dominant term in this case is the V_2 term.

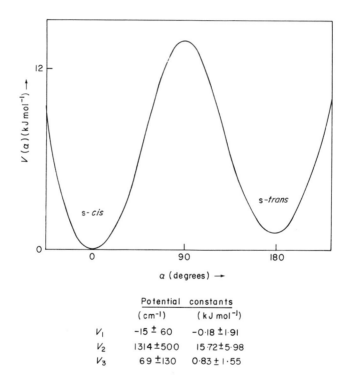

	Potential	constants
	(cm⁻¹)	(kJ mol⁻¹)
V_1	-15 ± 60	-0.18 ± 1.91
V_2	1314 ± 500	15.72 ± 5.98
V_3	69 ± 130	0.83 ± 1.55

Fig. 6.5   Potential energy function and constants for internal rotation in acrylic acid.

For the reasons already mentioned potential functions determined in this manner can, at best, be expected to be only a fair approximation to a "true" potential function for the internal rotation in a particular molecule. Improved accuracy in the data used contributes towards a remedy for this situation but, in general, a better description of the vibration can only be obtained by a more complete analysis of the dynamics of the internal rotation of the asymmetric internal top. The mathematical treatment of internal rotation in these cases is lengthy[3] and the details are beyond the scope of this book, however, an outline of the approach to the problem will be given and some of the more important consequences illustrated.

6.3 Internal Rotation in Completely Asymmetric Molecules

In the introduction to this chapter it was mentioned that the molecular dynamics of internal rotation in molecules with an asymmetric internal rotor are complicated by the fact that the inertial parameters of the molecule are functions of the internal rotation angle α. This makes the algebraic evaluation of the various parameters involved in the calculation of molecular energy levels, wave functions and potential functions very difficult and encourages a numerical approach to the problem.

The essence of the method is to express the inertial parameters of the molecule as Fourier series in the internal rotation angle. For example, the elements μ_{ij} of the inverse effective inertial tensor of the molecule are expressed in the form,

$$\mu_{ij} = \mu_{ij}(0) + \sum_n a_{ij}(n) \cos n\alpha + \sum_n b_{ij}(n) \sin n\alpha \qquad (6.4)$$

The μ_{ij} are determined numerically from a model for various values of the internal rotation angle by evaluating the elements of the effective inertial tensor at chosen values of α and inverting the corresponding matrices. If this procedure is carried out for a sufficiently large number of values of α in the range 0 to 360° the coefficients $a_{ij}(n)$ and $b_{ij}(n)$ of equation (6.4) may be determined. It is usually found that the Fourier series (6.4) converge fairly rapidly. The functional forms of other α dependent parameters are determined by a similar procedure. Once the Fourier coefficients of the various quantities are known the expectation values required for the computation of the molecular energy levels may be determined by using the appropriate basis functions as described in Chapter 2.

The formulation of the internal rotation problem follows the usual procedure for quantum mechanical calculations of this kind. It is first necessary to define a molecular model which can be used to determine the classical kinetic energy for a molecule. This is then transformed to the appropriate quantum mechanical form and the potential energy added to give the Hamiltonian for the system. The required energy levels and wave functions can then be determined by the usual methods. A suitable molecular model for

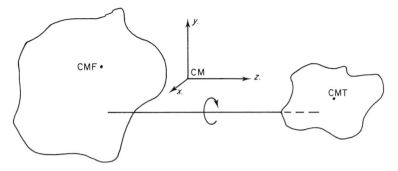

Fig. 6.6   An axis system for the development of the Hamiltonian for internal rotation in completely asymmetric molecules.

internal rotation in completely asymmetric molecules consists of two asymmetric, rigid parts joined by a rigid bond around which internal rotation can take place (Fig. 6.6). This is a particularly simple model in the sense that vibrational modes other than the torsion are neglected.

The axis system in which the kinetic energy is calculated is a matter of choice, some reference frames being more convenient than others. One suitable axis system is shown in Fig. 6.6. The origin coincides with the centre of mass of the molecule so the translational motion may be separated from other molecular motions and the z-axis is parallel to the axis of internal rotation (cf. the IAM method for symmetric internal tops in Chapter 5). The remaining two axes are fixed in some convenient manner with respect to the frame which may be chosen to be either of the two asymmetric groups in the molecule without affecting the problem. (The heavier group is often chosen to be the frame.) In this axis system the terms contributing to the total kinetic energy of the molecule fall into three categories, those associated with overall rotation, those associated with internal rotation of the top, and terms which couple the internal and overall rotational motions.

The corresponding quantum mechanical Hamiltonian may also be written as the sum of three parts,

$$H = H_R + H_T + H_{TR} \tag{6.5}$$

where H_R is a rotational Hamiltonian which is independent of the internal rotation angle, H_T is a function of α only and contains the potential energy $V(\alpha)$ and H_{TR} represents the coupling between the internal and overall rotation.

The energy levels and wave functions associated with equation (6.5) may be obtained from the energy matrix which can be set up using product basis functions of the type,

$$\Psi_{RT} = \psi_R \phi_T(\alpha) \tag{6.6}$$

where Ψ_{RT} is an eigenfunction of $H^0 = H_R + H_T$, ψ_R is an eigenfunction of H_R and $\phi_T(\alpha)$ an eigenfunction of H_T. The functions ψ_R may be either symmetric rotor wave functions or asymmetric rotor wave functions while the precise form of $\phi_T(\alpha)$ depends on the symmetry properties of the potential function $V(\alpha)$. As the problem is described here the energy matrix contains terms which connect different torsional states and torsional substates of the molecule and will be discussed further later. It is also a consequence of the choice of the axis system (Fig. 6.6) that H_{TR} is not necessarily small.

6.3.1 The torsional equation

The part of the Hamiltonian (6.5) which depends only on the torsional angle α has the form,

$$H_T = p^2(\alpha) F(\alpha) + F(\alpha) p^2(\alpha) + p(\alpha) F(\alpha) p(\alpha) + V(\alpha) \tag{6.7}$$

where $F(\alpha)$ is the reduced rotational constant for the torsional vibration and $p(\alpha)$ is the operator corresponding to the angular momentum of internal rotation. If certain terms which arise through the operation with the momentum operator on inertial parameters are neglected, then $V(\alpha)$ in equation (6.7) takes the usual form of a simple cosine potential function (equation (2.42)). The functional form of the reduced rotational constant, $F(\alpha)$, is determined from the molecular structure by an analogous procedure to that described earlier for obtaining the Fourier coefficients of the elements of the inverse effective inertial tensor.

The torsional energy levels and eigenfunctions can be obtained by setting up and diagonalizing a truncated version of the energy matrix of equation (6.7) using, for example, the free rotor wave functions as basis functions,

$$\phi(\alpha) = \frac{1}{\sqrt{2\pi}} e^{\pm im\alpha} \tag{6.8}$$

where m is an integer. The number of coefficients, V_n, which are used in such a calculation is, of course, governed by the available experimental data which relates to the shape of the potential function. The parameters, V_n, are varied in order to reproduce, as closely as possible, the observed torsional energy level patterns for the various rotamers and energy differences between the rotamers. The method of obtaining the corresponding eigenfunctions has previously been mentioned in Chapter 2.

The relationship between the torsional energy levels calculated using the Hamiltonian (6.7) and those for a pure threefold symmetric internal rotor has already been illustrated qualitatively in Fig. 6.2. In this diagram the energy level scheme (d) in fact shows the effect of the addition of only a single $F \cos \alpha$ term to the reduced rotational constant for the torsional vibration.

6.3.2 Rotational energy levels and torsion–rotation interactions

The rotational energy levels associated with the Hamiltonian (6.5) are eigenvalues of the energy matrix of the Hamiltonian H'',

$$H'' = H_R + H_{TR}$$

where H_R and H_{TR} have the form,

$$H_R = \mathbf{P}^+ \boldsymbol{\mu}(0)\, \mathbf{P}$$
$$H_{TR} = \mathbf{P}^+ \boldsymbol{\mu}(\alpha)\, \mathbf{P} - \sum_g G_g P_g \tag{6.9}$$

where $\boldsymbol{\mu}(0)$ and $\boldsymbol{\mu}(\alpha)$ are respectively torsionally independent and torsionally dependent parts of the inverse effective inertial tensor $\boldsymbol{\mu}$ for the molecule. $\mathbf{P}$ is the column vector $(P_x P_y P_z)$, the P_g being the quantum mechanical operators corresponding to the components of the total angular momentum in the molecule fixed axis system, and the G_g are functions of α dependent geometric terms and the internal angular momentum operator $p(\alpha)$.

It can now be seen why H_{TR} is not necessarily small in this axis system. The term $\boldsymbol{\mu}(0)$ in equation (6.9) is evaluated at the configuration of the molecule for which $\alpha = 0$ so the first term in the expression for H_{TR} is likely to be large. The second term contributing to H_{TR} is a Coriolis type term which couples the internal and overall angular momenta and in the axis system of Fig. 6.6 this too may be large.

In order to set up the energy matrix of H'' the functional forms of the elements of the inverse effective inertial tensor $\boldsymbol{\mu}$ and the G_g terms, must be determined by the procedure already described. The energy matrix can then be set up using the basis functions of equation (6.6) with the torsional eigen-functions obtained by solution of the torsional equation as described in Section 6.3.1. In this representation both H_R and H_T are diagonal in the tor-sional quantum number v, however H_{TR} contributes elements to the energy matrix of H'' which are off-diagonal in v. In particular, the off-diagonal elements arise from the Coriolis terms together with terms of the form $\langle \mu(\alpha)_{ij} \rangle_{vv}, P_i P_j$ (see the discussion in Section 2.9). Consequently the energy matrix of H'', which is diagonal in J, contains terms connecting different torsional states as well as elements connecting torsional sub-states of oppo-site symmetry within a particular block. The infinite energy matrix of H'', may be restricted to a manageable size by truncating the basis set at a point which allows the eigenvalues of interest to be determined to the required accu-racy. The energy matrix is usually further simplified by using a Van Vleck transformation to reduce to higher order the matrix elements connecting diffe-rent torsional states and those connecting states belongng to different rota-meric forms of the molecule. This procedure gives rise to an effective Hamiltonian for each rotamer which may be treated independently to give the rotational energy levels for the different rotamers. The retention in the effective Hamiltonian for a particular rotamer of the off-diagonal terms connecting torsional sub-states of opposite parity will account for any observable deviations of the rotational energy level pattern for the rotamer from that of a rigid rotor.

A consequence of setting up the molecular Hamiltonian in the axis system of Fig. 6.6 is that care must be taken when comparing observed and calculated rotational constants for individual rotamers that both sets of parameters are referred to the same axis system. When necessary the axes of the calculated parameters may be rotated to axes consistent with the observed parameters.

6.3.3 The Hamiltonian in the instantaneous principal axis system of the molecule

An alternative to setting up the Hamiltonian for internal rotation in an axis system in which one of the axes lies parallel to the axis of internal rotation is to use the instantaneous principal axes of the molecule thus ensuring that the inertial tensor of the molecule is always diagonal.[4] Such an axis system will therefore rotate as internal rotation occurs in the molecule. The main advan-tages of using this axis frame are that the off-diagonal elements of the type $\langle \mu(\alpha)_{ij} \rangle_{vv'} P_i P_j$ in the energy matrix of the Hamiltonian and the Coriolis-type terms are much smaller than in the axis system of Fig. 6.6 and the observed and calculated rotational constants may be compared in a quite straight-forward manner. Like the PAM for symmetric internal tops (Chapter 5) this axis frame seems to have its main application when the top is light as for a hydroxyl group.

6.3.4 The reduced potential

When only the height of the barrier hindering internal rotation for a molecule is required, it is possible to simplify the torsional equation (6.7) through the use of a reduced potential $V'(\theta)$. This involves writing the Hamiltonian (6.7) in such a way that the angular dependence of $F(\alpha)$ is removed and the potential function $V(\alpha)$ modified to take account of this change. The Hamiltonian then takes the simpler form,

$$H'_T = F'p^2(\theta) + V'(\theta) \qquad (6.10)$$

where F' is calculated from a molecular model and the relationship between θ and α is such that $\theta = \alpha$ at 0 and 2π radians.

The reduced potential determined in this way will have the same barrier heights as $V(\alpha)$ but the shape of the potential function and the values of the potential constants will differ somewhat from those of the "true" potential function.

6.3.5 Internal rotation in 3-fluoropropene

Some of the points mentioned in the preceding sections may become clearer in the context of a particular example. Internal rotation in 3-fluoropropene has been studied very extensively using the methods described above.[3d] Spectroscopic evidence shows that, in the gas phase, this molecule exists in the two rotameric forms shown in Fig. 6.7. The experimental data used in the derivation of the potential function for internal rotation of the CH_2F group is summarized in Table 6.3. Also included in this table is an indication of the source of the various data together with those parts of the potential function which are mainly dependent on a particular piece of data. Using the procedure outlined in Section 6.3.1 the first six potential constants V_1 through V_6 have been determined and are given in Fig. 6.8 which illustrates the shape of the associated potential function. Also included in Fig. 6.8 for comparison are the corresponding coefficients of the reduced potential function. It can be seen that these differ considerably from those of the "true" potential function

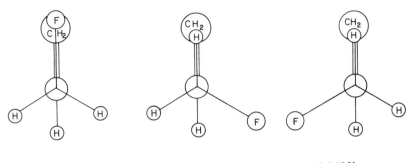

cis (a = 0)   *gauche* (α ≈ 120°, 240°)

Fig. 6.7   The stable forms of 3-fluoropropene.

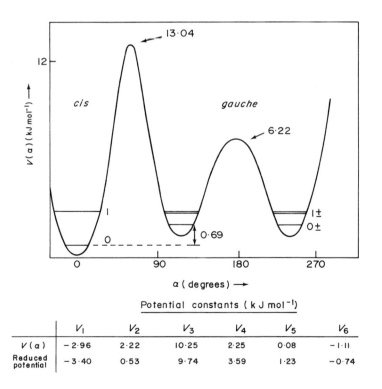

Potential constants (k J mol⁻¹)

	V_1	V_2	V_3	V_4	V_5	V_6
$V(\alpha)$	− 2·96	2·22	10·25	2·25	0·08	− 1·11
Reduced potential	− 3·40	0·53	9·74	3·59	1·23	− 0·74

Fig. 6.8   Experimental potential function and related potential constants for internal rotation in 3-fluoropropene.

Table 6.3   Experimental data used to obtain the potential function for internal rotation in 3-fluoropropene

Data	Source	Principal effect on the potential function
(1) Energy separations between the torsional ground states and excited states		
(a) *cis* $1 \leftarrow 0$ $164 \cdot 5 \pm 1 \text{ cm}^{-1}$ $2 \leftarrow 0$ $322 \cdot 5$ $3 \leftarrow 0$ $474 \cdot 0$ $4 \leftarrow 0$ $621 \cdot 0$	far-infrared frequencies	shape of *cis* well
(b) *gauche* $1^{\pm} \leftarrow 0^{\pm}$ $84 \cdot 6 \pm 2 \cdot 8 \text{ cm}^{-1}$ $2^{\pm} \leftarrow 0^{\pm}$ 185 ± 25	microwave intensities	shapes of the barrier at the *trans* position and the *gauche* well
(2) Torsional splittings for the *gauche* rotamer $0^{-} \leftarrow 0^{+}$ $0 \cdot 125 \pm 0 \cdot 015 \text{ MHz}$ $1^{-} \leftarrow 1^{+}$ $7 \cdot 0 \pm 0 \cdot 2$ $2^{-} \leftarrow 2^{+}$ $184 \cdot 8 \pm 3 \cdot 0$	microwave frequencies	
(3) $\Delta E \, (gauche-cis)$ $0 \cdot 69 \pm 0 \cdot 28 \text{ kJ mol}^{-1}$	microwave intensities	energy difference between *gauche* and *cis* minima.

although the *cis–gauche* and *gauche–gauche* barrier heights are identical with those of the "true" function as is the *cis–gauche* rotamer energy difference.

Apart from the inherent uncertainties in the experimental data of Table 6.3 there is a further important source of error in the determination of potential constants in this case. The detailed structures of the *cis* and *gauche* rotamers are such that a simple rotation around the C—C bond from the *cis* conformation does not produce exactly the established *gauche* structure. There is therefore some doubt as to which structure should be used in the determination of the functional form of $F(\alpha)$ for use in equation (6.7). This problem can be resolved by using a hybrid $F'(\alpha)$ derived from the functions obtained using each rotameric structure, the hybrid being constructed in such a way that it reproduces the correct value of *cis* $F(\alpha)$ at $\alpha = 0°$ and that of *gauche* $F(\alpha)$ at $\alpha = 120°$. This uncertainty of the precise form of $F(\alpha)$ for the molecule represents one of the main sources of error in the potential function shown in Fig. 6.8 its greatest effect being on the height of the *gauche–gauche* potential barrier.

The rotational energy levels of 3-fluoropropene are obtained from the torsion–rotation Hamiltonian corresponding to equation (6.9). In this case the simplification of the energy matrix by the Van Vleck transformation results in the removal of off-diagonal elements connecting blocks of nearly degenerate *gauche* levels to other nearly degenerate *gauche* blocks together with those matrix elements connecting *cis* energy levels to *gauche* blocks. After the Van Vleck transformation the effective Hamiltonians for the *cis* and *gauche* rotamers have the form,

$$H_{cis} = R_{xx}P_x^2 + R_{yy}P_y^2 + R_{zz}P_z^2 + R_{yz}(P_zP_y + P_yP_z)$$

$$H_{gauche}^{\pm} = \begin{bmatrix} R_{xx}^{(+)}P_x^2 + R_{yy}^{(+)}P_y^2 + R_{zz}^{(+)}P_z^2 & R_{xy}(P_xP_y + P_yP_x) + G_yP_y \\ \quad + R_{yz}^{(+)}(P_yP_z + P_zP_y) & \quad + R_{xz}(P_xP_z + P_zP_x) + G_zP_z \\ \hline iR_{xy}(P_xP_y + P_yP_x) + G_yP_y & R_{xx}^{(-)}P_x^2 + R_{yy}^{(-)}P_y^2 + R_{zz}^{(-)}P_z^2 \\ \quad + iR_{xz}(P_xP_z + P_zP_x) + G_zP_z & \quad + R_{yz}^{(-)}(P_yP_z + P_zP_y) + \Delta \end{bmatrix}$$

$$(6.11)$$

where Δ is the energy difference between the nearly degenerate *gauche* $\pm$ torsional sub-states and where the superscripts $\pm$ on the effective rotational constants refer to these same sub-states.

The energy matrices of the effective *cis* and *gauche* Hamiltonians can be diagonalized and the various parameters of expression (6.11) adjusted to provide the best fit to the observed rotational transition frequencies for each rotamer. The interpretation of the effective rotational constants obtained by these methods requires some care. In some cases these constants contain a considerable contribution from vibration–rotation interactions arising from the Van Vleck procedure and this can lead to deceptively large differences between rotational constants in different torsional sub-states for example. Consequently differences between rotational constants for, say the two lowest torsional sub-states of the *gauche* rotamer of a molecule like 3-fluoropropene do not necessarily reflect significant structural differences in these two states.

6.4 Internal Rotation in some Selected Molecules

Although experimental data on internal rotation of non-symmetric tops is less plentiful than for symmetric internal rotors a considerable number of examples has been studied. No attempt is made here to provide an exhaustive list of these compounds; rather the purpose is to select some examples which illustrate the nature of the data available and the varying depth of treatment of the problem. For the purposes of classification of the examples it is convenient to use the dominant term in the Fourier expansion for the potential energy curve.

6.4.1 Molecules with predominantly twofold barriers

A number of monosubstituted benzene derivatives have potential functions hindering internal rotation around the bond joining the substituent to the ring which are mainly V_2 in character. An example of this is nitrobenzene. The torsional frequency of the nitro group in this planar molecule has been determined from the inertial defect $(I_c - I_a - I_b)$ of the ground and first excited torsional states as 49 cm^{-1}. If a V_2 term only is included in the expression for the potential function hindering the torsion then the barrier hindering internal rotation in this molecule may be calculated from the appropriate expression in the form of equation (6.2) as $13 \cdot 5 \pm 6.8$ kJ mol^{-1}. The behaviour of the rotational constants as a function of torsional state is not, however, easy to reproduce and there is evidence that it is necessary to use a model which permits a synchronous variation of the ONO angle during the torsional vibration to obtain satisfactory agreement between the calculated and observed variation of rotational constants with torsional state. Analogous structural relaxation might be expected, on intuitive grounds, to affect the dynamics of internal rotation in many other cases. However, inclusion of these effects complicates the internal rotation problem and there may be ambiguity over the choice of the mode of relaxation, consequently analysis of internal motions which include structural relaxation has been applied in only a limited number of cases.

Internal rotation of the OH group in phenol is expected, from simple considerations, to be governed by a twofold symmetric potential function. Arguments based on the relative ease of delocalization of the "p-type" lone-pair in preference to the "sp^2-type" lone-pair of electrons on the oxygen atom predict planarity of the molecule in the ground state with significant double bond character in the C—O bond. These expectations are borne out by experiment, the infrared and microwave spectra in particular being consistent with a planar ground state. The effects of torsion–rotation interactions are apparent in the microwave spectrum of the molecule and in the spectra of various isotopically substituted species. Analyses of the data pertaining to the internal rotation problem assuming the OH group to behave as a symmetric top, show the barrier to internal rotation to be $14.0 \pm 0.5 \, \text{kJ mol}^{-1}$. In phenol OD the barrier apparently increases to $15.3 \, \text{kJ mol}^{-1}$. Such a variation due to isotopic substitution is rather unexpected and this large difference would seem to indicate that in this case neglected vibrational interactions probably have a significant role.

In benzoyl fluoride the variation of the rotational constants with torsional quantum number for the first seven torsional states is very nearly linear. This, together with the absence of any torsion–rotation splitting in any of the torsional states studied, points to a high twofold barrier to internal rotation. The V_2 barrier, calculated from an expression of the form of equation (6.10), i.e. ignoring any α dependence of the reduced rotational constant, is $18.6 \pm 1.9 \, \text{kJ mol}^{-1}$. This can be compared with the corresponding barrier height in benzaldehyde of $19.5 \, \text{kJ mol}^{-1}$, both values being consistent with the presence of some π bond character in the C—C bond to the aromatic ring in these compounds. In the case of benzaldehyde which is known to be planar in its ground state, there is evidence that vibration–internal rotation interactions may be contributing significantly to the observed variation of effective rotational parameters with torsional quantum number. Even the inclusion of α dependence of the reduced rotational constant fails to reproduce the observed variation of rotational constants with torsional state.

Similar effects to those described for benzaldehyde have been detected for nitrosobenzene. Once again the model with only a single internal degree of freedom and including the α dependence of the reduced rotational constant, fails to account satisfactorily for the observed data, presumably for reasons similar to those in the case of benzaldehyde. In fact for both benzaldehyde and nitrosobenzene there is reason to believe that only about 50% of the torsional dependence of the effective rotational parameters can be accounted for by internal rotation the remainder arising from interactions between the torsional mode and other low frequency vibrations. The calculated V_2 barrier for nitrosobenzene is 15.3 kJ mol^{-1}.

Amongst the simplest molecules exhibiting internal rotation are H_2O_2 and H_2S_2. In H_2O_2 the O—O stretching vibration occurs close to 900 cm^{-1} which is considerably higher than any expected frequency of the torsional vibration so the model with just one internal degree of freedom is expected to be applicable in this case. The form of the potential function hindering internal rotation is shown in Fig. 6.9 and the torsional energy levels are obtained by

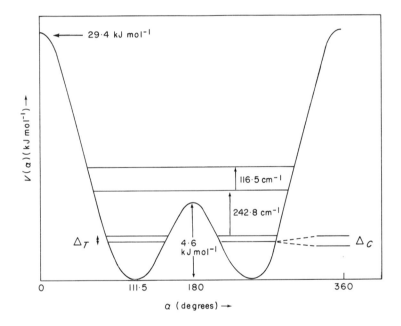

Fig. 6.9   Potential function for internal rotation in H_2O_2.

solving an equation analogous to (6.7).[5] The associated energy levels are shown schematically in Fig. 6.9. The large splitting, Δ_T, is due to tunnelling through the *trans* barrier and the smaller splitting, Δ_C, due to tunnelling through the higher *cis* barrier. The first three potential constants have been determined as $993, 636$ and 44 cm^{-1} from splittings of torsional energy levels determined from the far-infrared and millimetre wave spectra of the compound. The splitting Δ_T is 343 GHz and the minima of the potential function are at $111.5°$

from the *cis* conformation. The smaller splitting Δ_C, is very small and has not yet been resolved. The barriers hindering internal rotation at the *cis* and *trans* conformations are 29·4 and 4·6 kJ mol^{-1} respectively.

A similar analysis of spectral data for H_2S_2[6] reveals that the internal rotation splitting in the ground state is very much smaller than in H_2O_2, too small, in fact, to be resolved in the millimetre wave spectrum of the molecule by conventional techniques; Lamb Dip spectroscopy has shown the $0^- \leftarrow 0^+$ splitting to be only 60 kHz.[7] However, it is possible to resolve the splitting in the first excited state of the torsional vibration. The barrier to internal rotation at the *trans* conformation of H_2S_2 (28·4 kJ mol^{-1}) is therefore considerably higher than the corresponding barrier in H_2O_2. The barrier at the *cis* conformation (30·1 kJ mol^{-1}) is also higher than that for hydrogen peroxide. The increase in barrier heights is attributed mainly to an increase in the π-bond character in the S—S bond as compared to the O—O bond through hyperconjugation effects. Further support for this conclusion comes in the form of the much smaller equilibrium angle α_e of $\sim 90°$ obtained for H_2S_2 compared with the value of 111·5° for H_2O_2.

The existence of π-bond character in the C—C bonds of acrylic acid and acrylyl fluoride is thought to be mainly responsible for the dominance of the V_2 term in the potential functions for internal rotation in these molecules. Both are known to be planar in the ground state and conjugation analogous to that in butadiene gives rise to the two stable rotamers, s-*cis* and s-*trans* (Fig. 6.4). Spectroscopic studies in the gas phase confirm the existence of these rotamers. It might be expected that a similar situation would be found in acrolein. Ultraviolet spectroscopy of acrolein vapour reveals the presence of two rotamers, the s-*trans* and most probably the s-*cis* form, although it is possible that instead of s-*cis* the second rotamer may be *gauche*. The difference in energy between the rotamers is of the order of 8·4 kJ mol^{-1} so only about 4% of the molecules exist in the higher energy conformation at room temperature. This agrees well with conclusions recorded on the basis of ultrasonic studies in the condensed phase which suggest that any finite barrier between s-*cis* and s-*trans* acrolein will be of the order of 20·9 kJ mol^{-1}. Glyoxal also belongs to this general family of molecules and electronic spectroscopy again shows that two forms exist, the s-*trans* rotamer being more stable than the s-*cis* by about 13·4 kJ mol^{-1}. In this case the presence of the s-*cis* form has been confirmed by microwave spectroscopic studies and the potential constants $V_1 = 1125$ and $V_2 = 930$ cm^{-1} have been determined from the torsional frequency of the *trans* rotamer and the energy difference between the two stable rotamers.

Barriers to internal rotation in the region of 25 kJ mol^{-1} and higher have been determined for a great many molecules by nuclear magnetic resonance spectroscopy. Unfortunately in these cases detailed information about the potential function is not usually available. However, in molecules with these high barriers, effects of, for example, π-bonding in the bond around which internal rotation is occurring, or large steric contributions to the barrier hindering the rotation, are often apparent. Some examples of higher barrier cases are given in Fig. 6.10.

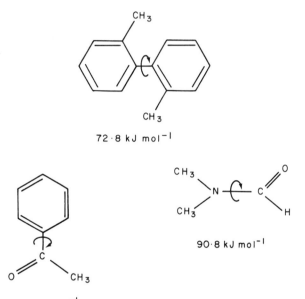

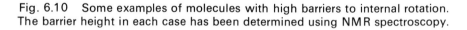

Fig. 6.10   Some examples of molecules with high barriers to internal rotation. The barrier height in each case has been determined using NMR spectroscopy.

6.4.2 Molecules with a predominantly threefold potential function

In the concluding section of Chapter 5 a brief reference was made to the effect of partial deuteration of the methyl group of acetaldehyde on internal rotation in the molecule. This case is rather a special one since the deuterated group, although no longer a symmetric internal rotor, is still associated with a threefold symmetric potential function similar to that for acetaldehyde itself. The barrier height, determined by the methods described in Sections 6.3.1– 6.3.3, is 4·9 kJ mol^{-1} for CH_2DCHO and 4·7 kJ mol^{-1} for CHD_2CHO.

These results are in satisfactory agreement with the value of $4{\cdot}9\,kJ\,mol^{-1}$ for acetaldehyde itself.

An interesting class of compounds in which internal rotation has been investigated quite extensively includes some simple alcohols and their sulphur analogues the thiols. Many molecules containing an OH group can be considered to be derivatives of methanol in which one, or more, of the hydrogen atoms of the methyl group has been replaced by another atom or group. The potential function for internal rotation around the C—O bond therefore loses the threefold symmetry characteristic of methanol itself. The rotamers most likely to exist are shown in Fig. 6.11 the conformer labelled I being conventionally referred to as the *trans* rotamer and the conformers II and III, clearly energetically equivalent when $X = Y$, as the *gauche* rotamers. A similar situation exists for thiols although experience has shown that it is unwise to assume that the relative stabilities of the conformers is necessarily the same as for the corresponding alcohols.

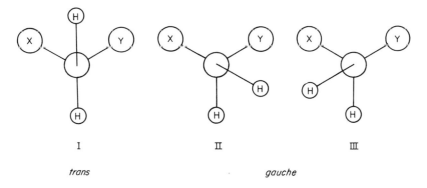

Fig. 6.11   Rotameric forms of a molecule of the type CHXY—OH.

Some results of spectroscopic investigations of alcohols and thiols in the gas phase are summarized in Chapter 4 in Table 4.2 in the form of the most favoured conformers found in each case. It appears that in cases where the *trans* conformer is present in significant concentration the *gauche* conformer is present also. However, when the *gauche* conformer is found to be energetically the more favoured the *trans* conformer is seldom, if ever, present in any abundance.

Hydroxyacetonitrile is one of the most thoroughly investigated alcohols from the point of view of internal rotation.[8] The microwave spectra of the parent molecule and the OD analogue show that only the *gauche* rotamer is present in significant abundance in the gas phase and that the separations between the sub-states $0^{\pm}$ of the OH and OD species are 110·7 and 16·7 GHz respectively. Deviations of the spectra of both isotopic species from that of a rigid rotor are, to a large extent, satisfactorily accounted for using the theory described in Sections 6.3.1–6.3.2.

Unfortunately only two pieces of experimental data are available for the determination of the potential function hindering the torsion of the OH group. However, if the value of the equilibrium angle, α_e, of the *gauche* rotamer is fixed, a cosine function can be determined which will reproduce the observed $0^- \leftarrow 0^+$ energy level separations of the OH and OD isotopic species. The most acceptable potential functions for hydroxyacetonitrile are obtained for the range $106 < \alpha_e < 116°$. When α_e is greater than 116° a deep minimum containing bound energy levels appears at the *trans* position, a fact which cannot be reconciled with the absence of *trans* rotamer rotational transitions in the microwave spectrum. Potential functions determined with $\alpha_e < 106°$ have a positive V_3 term which is unacceptable. A potential function which consists of cos α + cos 2α terms only, in phase, has the appearance of a typical, symmetrical double minimum function. The addition of a sizeable cos 3α term tends to destroy this characteristic shape and to introduce minima at positions with the hydroxyl hydrogen atom eclipsing the hydrogens of the CH_2 group in a manner contrary to that observed for methanol. The subtraction of such a term merely accentuates the minima and alters the separation between them, simultaneously increasing the barrier height. If the potential function is to retain a physically reasonable form, the V_3 term for hydroxyacetonitrile is expected to be negative, zero or, at most, small and positive. Some suitable potential functions are illustrated in Fig. 6.12 together with the corresponding Fourier coefficients. In each case the barrier at the *cis* position is relatively well determined when compared with that at the *trans* position. This results from the fact that the experimental data (the $0^- \leftarrow 0^+$ energy difference) is mainly determined by the nature of the *cis* barrier. Comparison of the observed differences in the rotational constants for the 0^+ and 0^- sub-states and those calculated using the eigenvectors of the torsional Hamiltonian for values of α_e in the range defined above, indicate that the best agreement is to be found for α_e closer to 116° than 106°.

The arguments which have just been outlined for hydroxyacetonitrile illustrate the rather indirect way in which a potential function has to be determined when only very limited data is available and also clearly demonstrates the subsequent uncertainty in the final potential function. Potential functions for propargyl alcohol, propargyl mercaptan and cyclopropanol have all been determined from similar limited data and are consequently

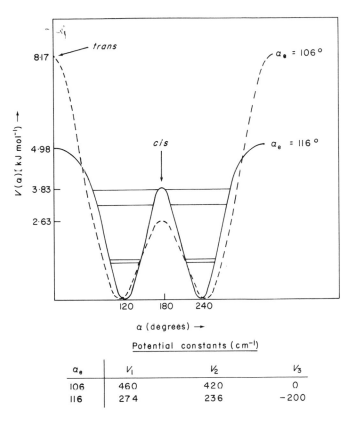

Fig. 6.12   Potential functions and Fourier coefficients for internal rotation in hydroxyacetonitrile.

subject to the same considerable uncertainties already discussed for hydroxy-acetonitrile. The *gauche–gauche* potential barriers for these compounds are summarized in Table 4.2 although in these cases the OH group has been treated as a symmetric top internal rotor.

For those alcohols and thiols where *trans* and *gauche* rotamers are stable in the gas phase the tunnelling splittings of the lowest torsional state of the *gauche* rotamers and the *gauche–gauche* barriers to internal rotation are summarized in Table 4.2 where they are known. In these cases the barriers to interconversion of the two *gauche* rotamers have been calculated assuming that the V_3 term dominates in the potential function for internal rotation of the hydroxyl group, an assumption which must await future investigations for its justification.

Further Reading and References

The analysis of internal rotation of asymmetric groups is relatively new and at present there are no textbooks dealing with this subject. For further insight into the treatment the reader is directed to ref. 3.

1. G. L. Carlson, W. G. Fateley and J. Hiraishi. *J. Mol. Structure*, **6**, 101 (1970).
2. K. Bolton, D. G. Lister and J. Sheridan. *J. Chem. Soc. Faraday. Trans. II*, **70**, 113 (1974).
3. (a) C. R. Quade and C. C. Lin. *J. Chem. Phys.* **38**, 540 (1963).
   (b) C. R. Quade. *J. Chem. Phys.* **47**, 1073 (1967).
   (c) J. V. Knopp and C. R. Quade. *J. Chem. Phys.* **48**, 3317 (1968).
   (d) P. Meakin, D. O. Harris and E. Hirota. *J. Chem. Phys.* **51**, 3775 (1969).
4. A. M. Mirri, F. Scappini, R. Cervellati and P. G. Favero, *J. Mol. Spectroscopy*, **63**, 509 (1976).
5. R. M. Hunt, R. A. Leacock, C. W. Peters and K. T. Hecht. *J. Chem. Phys.* **42**, 1931 (1965).
6. G. Winnewisser, M. Winnewisser and W. Gordy. *J. Chem. Phys.* **49**, 3465 (1968).
7. Y. Morino and S. Saito. *In* "Molecular Spectroscopy: Modern Research" (K. N. Rao and C. W. Matthews, Ed.) Chapter 2.1, London and New York, Academic Press, 1972.
8. G. Cazzoli, D. G. Lister and A. M. Mirri. *J. Chem. Soc. Faraday Trans. II*, **69**, 569 (1973).

7
Inversion

7.1 Introduction

The mode of vibration known as inversion exists, in principle, for all non-planar molecular species. Like internal rotation, it is a hindered internal motion which, in the classical sense, is associated with the interconversion of molecular configurations. The inversion vibration usually involves the interchange of two energetically equivalent configurations via a planar intermediate (Fig. 7.1(a)). A graph of the potential energy of the molecule against some suitable coordinate which describes the classical motion of the atoms, characteristically exhibits two minima separated by a potential energy barrier at the planar configuration, Fig. 7.1(b). This diagram represents a section through what, in general, is a multidimensional potential energy surface. A similar potential function is associated with certain ring puckering vibrations and these are discussed in Chapter 8.

From consideration of Fig. 7.1(b) it might be expected that only those molecules which attain sufficient vibrational energy to surmount the potential barrier can interchange between configurations I and III. However, a careful examination of experimental evidence, mainly spectroscopic in origin, suggests that this is not the case but that molecules which do not possess the requisite amount of vibrational energy for the classical motion to occur still appear to succeed in inverting. This is another example of "tunnelling", a process which has no classical counterpart and which can only be adequately accounted for by detailed consideration of the quantum mechanics of the inversion vibration.

The quantum mechanical characteristics of the inversion mode for a particular molecular species are determined to a considerable extent by detailed features of the potential function such as the barrier height and the

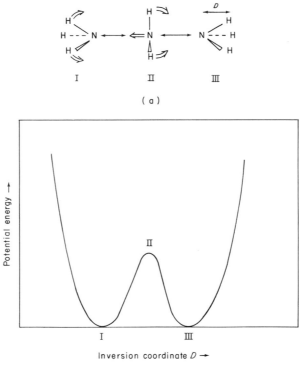

Fig. 7.1   Classical inversion in ammonia: (a) illustrates the classical inversion of the pyramidal configuration at the nitrogen atom. I and III are the stable configurations and II is the unstable planar form of the molecule; (b) shows the corresponding double minimum potential function for the vibration. A suitable inversion coordinate might be the distance of the nitrogen atom from the H_3 plane, for example.

separation of the two minima. In cases where the central potential barrier is extremely high, and this is true for most non-planar molecules, no inter-conversion between the stable configurations takes place on a reasonable time scale and so, for most purposes, the inversion vibration can be neglected. However, when the barrier to inversion is finite, tunnelling may become extremely important and the spectroscopic properties of the molecule can be dramatically affected, particularly when the atoms involved in the motion are light. Conversely, careful analysis of spectra of inverting molecules can yield data, in varying degrees of detail, about the potential function governing inversion in a particular case. For barriers greater than about 25 kJ mol^{-1} nuclear magnetic resonance spectroscopy has proved especially useful for determining barrier heights, and inversion has been studied by this method for many molecules where the groups involved in the vibration are heavy. However, much of the detailed information about the nature of the inversion vibration and its associated potential function has come from studies on molecules for which the barrier lies below 25 kJ mol^{-1} using microwave or infrared spectroscopic techniques. Quite a number of molecules in this latter category have been investigated, the oldest, and probably the best known example, being ammonia. In ammonia the barrier hindering the "umbrella like" inversion motion is 5·8 kJ mol^{-1} and in fact the first example of a spectroscopic transition to be observed in the microwave region of the spectrum was associated with this inversion.

Much of the discussion in this chapter centres around the inversion of the nitrogen pyramid in ammonia and substituted ammonias simply because these are amongst the most thoroughly studied examples. However, it should be remembered that studies have been carried out on inversion at centres other than nitrogen, and that the phenomenon has been detected even for relatively complicated high barrier cases such as cyclic carbanions and oxonium salts as well as for simpler phosphorous and arsenic pyramids.

7.2 Quantum Mechanical Characteristics of the Inversion Vibration

In the introduction it was pointed out that in order to satisfactorily explain the observed spectroscopic properties of a molecule for which inversion is important it is necessary to examine in detail the quantum mechanics of the vibration. Basically, this involves the determination of vibrational wave functions and energy levels in a double minimum potential function of the type shown in Fig. 7.1 by solving the Schrödinger equation for the inversion vibration (see Chapter 2).

A conveniently simple species to consider in order to illustrate the properties of inversion wave functions and energy levels is the symmetric top XY_3. If the rotation of the molecule and any interactions with other vibrations are neglected, the inversion motion in this case is relatively uncomplicated. The vibration can be described by a single coordinate (q) related to the structure of the molecule during inversion and the vibrational wave functions and energy levels may be obtained by solving a one-dimensional Schrödinger equation with the Hamiltonian,

$$H = -\frac{\hbar^2}{2\mu}\frac{d^2}{dq^2} + V(q) \tag{7.1}$$

where μ is the reduced mass for inversion and depends on the paths followed by the nuclei, and $V(q)$ is a double minimum potential function. Two possible classical paths for the nuclei are shown in Fig. 7.2 the corresponding reduced masses being

$$\mu(a) = \frac{3mM}{3m + M}$$

and

$$\mu(b) = \mu(a)\left(1 + \frac{3m}{M}\sin^2\theta\right)$$

(7.2)

where m and M are the masses of atoms Y and X respectively and θ is the equilibrium Y—X—Y angle. Neither of the paths depicted in Fig. 7.2 adequately describes the actual motion but rather they should be regarded simply as possible limiting cases.

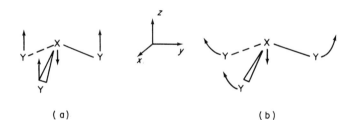

(a)   (b)

Fig. 7.2   Classical paths for inversion in XY_3 type molecules. (a) The atoms travel along a path which maintains a constant X—Y bond length. (b) The Y—Y distance is constant throughout the motion. The reference axes would normally have their origin at the centre of mass of the molecule.

Before considering any particular forms of $V(q)$ which might be used in equation (7.1) it is useful to examine some general features of the wave functions and energy levels of a molecule undergoing inversion.

7.2.1 Properties of molecular wave functions for an inverting molecule

The total energy, or more precisely, the Hamiltonian for a molecule is invariant under certain transformations of the co-ordinates of the nuclei and electrons making up the molecule. In the case of the non-planar XY_3 species one such transformation is the change of sign of the coordinate (e.g.

z in Fig. 7.2) which leads to inversion of the molecule. Inversion used in this sense therefore resembles a reflection in the Y_3 plane of the molecule and should not be confused with the term when it is used in the sense of the symmetry operation implying the coordinate transformation $x \rightarrow -x$, $y \rightarrow -y$, $z \rightarrow -z$. A consequence of the invariance of the energy under the inversion operation is that the wave function associated with a particular non-degenerate energy state of the molecule can, at most, change sign under the operation, since two successive applications of this twofold transformation must necessarily return the molecule to its original state. A wave function which changes sign is said to be antisymmetric with respect to inversion and one which retains its sign is symmetric.

The total wave function for a molecule (ψ_T) may be written as a product of the electronic (ψ_e), vibrational (ψ_v), rotational (ψ_r) and nuclear spin (ψ_s) wave functions and so the overall behaviour of ψ_T under inversion is governed by the symmetry of each individual component. For the present it may be assumed that all molecules are in the ground electronic state which is always symmetric and so only the remaining three parts of ψ_T need be considered. This will be done for the molecule XY_3 for the two limiting cases, where the barrier to inversion is infinitely high or negligibly small (planar molecule), and also for the case where the barrier is finite.

In the planar XY_3 molecule the vibration of interest is that in which the atom X oscillates perpendicularly to the plane of the three Y atoms. For small displacements, this vibration can be considered to be harmonic and a suitable wave function has the form,[1]

$$\psi_v(q) = N_v e^{-\frac{1}{2}q^2} H_v(\beta)$$

where N_v is a constant and β is proportional to q. The $H_v(\beta)$ are Hermite polynomials of degree v and involve only even powers of q when v is even and odd powers of q when v is odd. The transformation properties of $\psi_v(q)$ under inversion are therefore,

$$v \text{ even} \qquad \psi_v(q) \rightarrow (+1)\psi_v(q)$$
$$v \text{ odd} \qquad \psi_v(q) \rightarrow (-1)\psi_v(q)$$

The ground vibrational state ($v = 0$) is clearly symmetric and the first excited vibrational state ($v = 1$) antisymmetric as shown in Fig. 7.3(a). Similarly the symmetries with respect to inversion of the rotational wave functions for the molecule can be shown to depend on the K quantum number in the following manner,[2]

$$K \text{ even} \quad \psi_r(JKM) \to (+1)\psi_r(JKM)$$

$$K \text{ odd} \quad \psi_r(JKM) \to (-1)\psi_r(JKM)$$

Consequently, in the ground vibrational state of an inverting symmetric top molecule those states with K even are symmetric with respect to inversion while those with K odd are antisymmetric. The converse is true in the first excited vibrational state. Consideration of the symmetry properties of the nuclear spin wave functions ψ_s reveals that they are connected with the statistical weights and energy level populations and so they will have a considerable effect on the intensities of spectral lines. This point will be discussed further in connection with the inversion of the nonplanar XY_3 molecule.

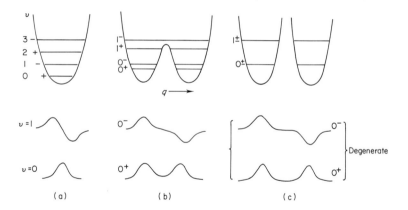

Fig. 7.3   Potential functions, energy levels and wave functions for (a) negligibly small, (b) finite, and (c) infinite barriers to inversion.

The non-planar XY_3 species with an infinitely high barrier to inversion has the same symmetry properties with respect to inversion as the planar molecule discussed above. The potential function for the vibration in which the X atom moves along a line perpendicular to the Y_3 plane now consists of two, essentially independent, wells each of which corresponds closely to the potential function for the nearly harmonic vibration of the atom X in the separate (1) and (2) configurations shown in Fig. 7.4. The wave functions for the entire system may now be expressed as linear combinations of the harmonic oscillator wave functions for the (1) and (2) configurations. If, for simplicity, only the two lowest states are considered then,

$$\psi = N(\psi_{(1)} \pm \psi_{(2)})$$

where N is a normalizing factor. Since the energies of the levels corresponding to $\psi_{(1)}$ and $\psi_{(2)}$ are identical, the combinations $N(\psi_{(1)} + \psi_{(2)})$ and $N(\psi_{(1)} - \psi_{(2)})$ in the infinite barrier limit are associated with degenerate energy levels. The potential function, energy levels and the wave functions when $v = 0$ are illustrated in Fig. 7.3(c). Diagrams showing the variation of the square of the wave functions with q indicate that the probability of finding the atom X in the plane of the Y atoms is exceedingly small, the vibrations of X being almost entirely confined to one or other of the separate wells.

(1)   (2)

Fig. 7.4   Separate configurations of the XY_3 species in the presence of an infinitely high barrier to inversion.

As the barrier to inversion is lowered from the infinite limit the extent to which $\psi_{(1)}$ and $\psi_{(2)}$ can interact increases and the energy levels of the system are modified. The extent of the modification can be seen if the Schrödinger equation for the problem is solved[3] using the variation method and the linear combination of harmonic oscillator wave functions described above. The energy levels are given by,

$$E = \frac{H_{(1)(1)} \pm H_{(1)(2)}}{1 \pm S_{(1)(2)}}$$

where $H_{(1)(1)} = H_{(2)(2)} = \int \psi_{(1)} H \psi_{(1)} \, d\tau$ $H_{(1)(2)} = \int \psi_{(1)} H \psi_{(2)} \, d\tau$ and $S_{(1)(2)} = \int \psi_{(1)} \psi_{(2)} \, d\tau$ and the associated wave functions are,

$$\psi = \frac{1}{\sqrt{2 \pm 2S_{(1)(2)}}} (\psi_{(1)} \pm \psi_{(2)})$$

When the inversion barrier is finite, the degeneracy of the energy levels in the limit of the infinite barrier is lifted (see Fig. 7.3) to an extent governed by $H_{(1)(2)}$ and $S_{(1)(2)}$. Evaluation of these integrals reveals that as the barrier is lowered so the energy level splitting increases until, when the barrier is zero, the pattern corresponds to that shown in Fig. 7.3(a) for the planar molecule. The energy level pattern typical of an XY_3 type molecule with an intermediate barrier height is shown in Fig. 7.3(b) together with the symmetries of the wave functions with respect to inversion. The qualitative appearance of the squares of the wave functions for $v = 0$ plotted against q for a typical case show that now there is a quite definite probability of finding the atom X in the plane of the Y atoms even in the very lowest vibrational states. The frequency corresponding to the energy separation between the two lowest inversion states of such a molecule (0^+ and 0^- as shown in Fig. 7.3) is often referred to as the "inversion frequency" and spectroscopic transitions between these states are termed "inversion transitions".

It has already been mentioned that the nuclear spin wave functions affect the statistical weights of the molecular states. The way in which this affects an inverting, non-planar XY_3 species is most readily discussed using the terminology of group theory.[4] For the purposes of illustration it is assumed that each Y atom has a nuclear spin of $\frac{1}{2}$ (e.g. a proton) and that the XY_3 molecule has C_{3v} symmetry in the non-planar equilibrium configuration. In the absence of inversion the vibration–rotation wave functions of such a molecule would normally be classified using the symmetry species of the C_{3v} group. However, in cases where inversion is important and all the possible operations, including inversion, which involve the permutation of the positions and spins of identical nuclei are considered, it becomes necessary to classify the molecule under a larger group which contains all the elements of the C_{3v} group together with those which convert the two C_{3v} configurations into each other. In the case of the XY_3 species considered here this larger group is isomorphous with D_{3h}. In these circumstances the rotation-vibration ($\psi_r \psi_v$) states originally classified under the C_{3v} group have a sub-structure which may be expressed in terms of the symmetry species of D_{3h} according to the correlation,

$$
\begin{array}{cc}
\underline{C_{3v}} & \underline{D_{3h}} \\
A_1 & \rightarrow A_1' + A_2'' \\
A_2 & \rightarrow A_2' + A_1'' \\
E & \rightarrow E' + E''
\end{array}
$$

So in instances where the inversion splitting is resolved, each rotation–inversion level of the molecule splits into two sub-levels with the symmetries given above.

An analysis analogous to that outlined in Section 5.3.2 reveals that the nuclear spin wave functions of the XY_3 molecule belong to the representation,

$$\psi_s = 4A_1' + 2E'$$

where the symmetry species have been assigned the labels commonly used in the D_{3h} group and use has been made of the fact that the spin wave functions are invariant under the inversion operation. The overall symmetries of the wave functions of the inverting XY_3 molecule, which are the direct products of $(\psi_v \psi_r)$ and ψ_s, are given in Table 7.1. The exclusion principle dictates that

Table 7.1   Symmetry classification of the wavefunctions for an inverting XH_3 type molecule and the statistical weights of the states

C_{3v}	$\psi_T = \psi_v \psi_r \psi_s$	Statistical weights
A_1	$\begin{cases} 4A_1' + 2E' \\ 4A_2'' + 2E' \end{cases}$	0 4
A_2	$\begin{cases} 4A_2' + 2E' \\ 4A_1'' + 2E'' \end{cases}$	4 0
E	$\begin{cases} 2A_1' + 2A_2' + 6E' \\ 2A_1'' + 2A_2'' + 6E'' \end{cases}$	2 2

in this case ψ_T should have A_2' or A_2'' symmetry therefore the statistical weights of the states are as shown in the table. It will also be noticed that certain rotational levels will not occur for some inversion states, a fact which is clearly exemplified in the spectrum of ammonia (see Section 7.2.3).

Many molecules for which the inversion barriers come into one of the three categories discussed above have been studied. Boron trifluoride is an example of a planar molecule for which the energy levels for the out-of-plane vibrational mode are nearly harmonic. For ammonia the potential barrier is moderately high ($5.8\ \text{kJ mol}^{-1}$) and in this case the $0^- - 0^+$ energy separation is $0.79\ \text{cm}^{-1}$. On the other hand, NF_3, AsH_3 and PH_3 have extremely high barriers so the inversion splitting is very small, and the molecules have little chance of inverting on a reasonable time scale. However, it should be noted that even for unlikely cases like AsH_3 for example, there is a possibility of inversion occurring given a time scale of the order of years.

7.2.2 Potential functions for the inversion mode

The detailed solution of equation (7.1) for an inverting molecule depends on the form of the expression used to describe the potential function for the vibration. Consequently, there have been many attempts to establish the most satisfactory form for this expression, the suitability of particular potential functions often being judged by the level of success achieved in predicting the ammonia inversion spectrum. The problem was first studied by Hund as early as 1927 and since then many potential energy functions have been tried, however, only certain selected examples will be considered here.

7.2.2.1 Harmonic oscillator perturbed by a Gaussian barrier

One of the most convenient ways of constructing a double minimum potential function for a vibration such as inversion is by superimposing a Gaussian function on the harmonic oscillator potential function,[5]

$$V(q) = \tfrac{1}{2}aq^2 + b\,e^{-c^2q^2} \tag{7.3}$$

where a, b and c are constants. The function $V(q)$ is characterized by three independent parameters so three experimentally observed quantities are required to define it in a particular case. Ideally these would be the separations between different inversion states of the molecule itself or of a suitable isotopic modification. In any particular case the variable parameters of equation (7.3) are adjusted in order to match the calculated energy level separations with the observed pattern.

The vibrational energy levels associated with $V(q)$ are obtained by solving a one dimensional Schrödinger equation using the Hamiltonian,

$$H = -\frac{\hbar^2}{2\mu}\frac{d^2}{dq^2} + \tfrac{1}{2}aq^2 + b\,e^{-c^2q^2} \tag{7.4}$$

This expression can conveniently be written in the form,

$$H = H^0 + H^1$$

where H^1 is the last term in equation (7.4) and may be regarded as a perturbation on the first two terms which correspond to the Hamiltonian of the harmonic oscillator H^0. The energy levels associated with the Hamiltonian may be determined by the methods outlined in Chapter 2, using the harmonic oscillator wave functions as basis functions. The behaviour of the rotational constants as functions of the vibrational quantum number for such a potential is also discussed in Chapter 2. In instances where the large degree of anharmonicity of the vibration prevents an observed energy level pattern from being satisfactorily reproduced, a quartic term of the type dq^4 may be added to equation (7.3) provided sufficient experimental data is available.

Table 7.2   Comparison of the observed and calculated inversion energy levels for ammonia.

State	Observed/cm^{-1} (i.r.)	Calculated/cm^{-1}	
		(HO + Gaussian)^a	(Manning)^b
0^+	0·0	0·0	0·0
0^-	0·79	0·78	0·83
1^+	932·4	932·4	935
1^-	968·2	968·8	961
2^+	1602	1603	1610
2^-	1882·2	1884	1870
3^+	2383·5	2387	2360
3^-	2895·5	2895	—
4^+	3442	3457	—

^{a} (HO + Gaussian) $E(0^+) = 513\cdot5$ cm^{-1}; barrier height $= 2031$ cm^{-1}; $q_m = 38\cdot6$ pm
^{b} (Manning)   barrier height $= 2076$ cm^{-1}; $q_m = 37\cdot0$ pm
q_m is half the separation between the potential minima.

Some of the parameters characterizing a double minimum function of this type for ammonia are given in Table 7.2, together with a comparison between the calculated energies of the inversion levels relative to the ground state (0^+) and those determined experimentally by infrared spectroscopy. In the case of the ammonia spectrum no significant improvement to the fit given in Table 7.2 is obtained by introducing a quartic term into the potential function.

G

7.2.2.2 Quadratic-quartic potential function

Another useful way of expressing the potential energy function for a large amplitude vibration is,[6]

$$V(q) = \sum_v a_v q^v \qquad (7.5)$$

A suitable form of equation (7.5) for the double minimum potential function governing the inversion mode is,

$$V(q) = aq^2 + bq^4 \qquad (7.6)$$

where the constant a has a negative sign.

Using this type of potential function the inversion energy levels are most readily obtained by transforming the coordinate q and its conjugate momentum in the appropriate Hamiltonian (expression (7.1)) into dimensionless space. Several possible transformations may be employed, a convenient one being defined by,

$$\xi = (\mu\gamma)^{\frac{1}{2}}q$$
$$\gamma = 4\pi^2 v_0/h$$

where μ is the reduced mass and v_0 is a scale factor with the dimension of frequency. In terms of the new coordinate, ξ, the reduced Hamiltonian has the form,

$$H = \tfrac{1}{2}hv_0[P^2 - \xi^2 + v\xi^4]$$

where $P = i^{-1}(d/d\xi)$ and where the relations,

$$a = -\tfrac{1}{2}hv_0(\mu\gamma)$$
$$b = \tfrac{1}{2}hv_0(\mu\gamma)^2 v$$

have been used. The energy levels and wave functions of this reduced Hamiltonian may be calculated with the harmonic oscillator wave functions as basis functions using the method outlined in Chapter 2.

7.2.2.3 Manning potential function

One of the most successful of the earlier potential functions for the inversion vibration was devised by Manning as long ago as 1935. This has the form,[7]

$$V(q) = \frac{1}{\alpha}\left[-C\,\mathrm{sech}^2\frac{q}{2\rho} - D\left(\mathrm{sech}^2\frac{q}{2\rho} - \mathrm{sech}^4\frac{q}{2\rho}\right) \right] \qquad (7.7)$$

where C, D, ρ and α are all constants and $\alpha = k\rho^2$ where k is proportional to the reduced mass for the vibration. The Schrödinger equation can be solved exactly using equation (7.7), the constants being adjusted to reproduce the observed energy level pattern. The energy levels of ammonia calculated using this function have been listed in Table 7.2 for comparison with the harmonic oscillator–Gaussian barrier function. This potential function has now largely been superseded by the computationally more convenient and probably more representative functions of the type described above.

7.2.3 Rotation inversion spectrum of the XY_3 symmetric top

Until now we have considered mainly the vibrational energy levels of an inverting molecule. However, associated with each vibrational state, there is a manifold of energy levels arising from the rotational motion of the molecule. The rotational energy levels of symmetric rotors have already been described in Section 2.8.3 and the fact that the number of possible spectroscopic transitions between the various energy levels of a molecule is limited by certain selection rules has already been noted.

The intensity of a particular rotation–inversion transition is related to the square of the transition moment integral (S) which is given for a totally symmetric electronic state by:

$$S = \int \psi_r^* \psi_v^* \mu \psi_{r'} \psi_{v'} \, d\tau \qquad (7.8)$$

where μ is the dipole moment operator. This expression will be non-zero provided the integrand is totally symmetric with respect to inversion. Whether or not this is the case for a given set of r and v subscript values may be established by considering the product of the individual symmetries of $(\psi_r^* \psi_v^*)$, μ and $(\psi_{r'} \psi_{v'})$. For the XY_3 species, μ lies along the symmetry axis of the molecule and inversion causes the direction of μ to be reversed. The operator is thus antisymmetric with respect to inversion so for equation (7.8) to be non-zero the symmetries of $(\psi_r^* \psi_v^*)$ and $(\psi_{r'} \psi_{v'})$ must be opposed.

G*

The rotational selection rules for a symmetric rotor are $\Delta J = 0, \pm 1$ and $\Delta K = 0$ and the latter condition dictates that ψ_r^* and $\psi_{r'}$ in equation (7.8) will have the same parity (see Section 2.1). Consequently, ψ_v^* and $\psi_{v'}$ must have opposite symmetries with respect to inversion, giving a vibrational selection rule which allows transitions only between inversion states of opposite parity.

The classic example of such a spectrum is the ammonia spectrum where, as can be seen from Table 7.2, the pure inversion transition $0^- \leftarrow 0^+$ occurs at 0.79 cm^{-1} in the microwave region of the spectrum while the vibrational transitions $1^\pm \leftarrow 0^\mp$ lie in the infrared region near 950 cm^{-1}. Because of the very large value of the rotational constant (B_0) for ammonia (~ 10 cm^{-1}) transitions connecting the rotational states of the 0^+ inversion level with those of the 0^- level (and obeying the $\Delta J = \pm 1, \Delta K = 0$ selection rules) are found in the far-infrared region whilst those with $\Delta J = 0, \Delta K = 0$ (the pure inversion transitions) occur in the microwave region of the spectrum. This is summarized in Fig. 7.5 where it will be noticed that some levels with $K = 0$ are shown as dashed lines. These levels in fact do not exist for NH_3 as a consequence of the required antisymmetry of the total wave function with respect to interchange of any two protons (Section 7.2.1). Careful examination of the pure inversion transitions reveals a dependence on the rotational quantum numbers which may be written in the form,[8]

$$v = v_0 \exp[aJ(J + 1) + bK^2 + cJ^2(J + 1)^2 + \ldots]$$

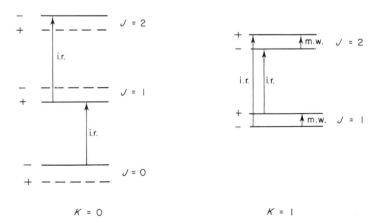

Fig. 7.5   Transitions in the inverting ammonia molecule. The labels i.r. and m.w. indicate whether a transition lies in the far-infrared or microwave region of the spectrum.

This type of expression is found to fit all but the $K = 3$ transitions quite accurately. The failure for $K = 3$ lines is due to a further rotation–vibration interaction which lifts the K degeneracy in levels where K is a multiple of three. In NH_3 the Pauli exclusion principle again causes one of the levels of each K doublet to vanish and so only a single displaced transition is observed rather than a doublet. This effect is largest for $K = 3$ and becomes progressively smaller for $K = 6, 9$, etc.

7.3 Inversion in Asymmetric Rotors

In many asymmetric rotors the potential function governing the inversion vibration retains the twofold symmetry properties of the functions described previously for a symmetric rotor. Consequently, the general comments on the vibrational energy level pattern in a double minimum potential which were made in Section 7.2.1 are often applicable to asymmetric top molecules.

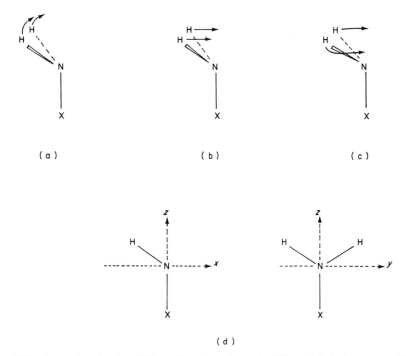

Fig. 7.6   Inversion in the NH_2—X molecule. (a), (b) and (c) show possible classical pathways for the hydrogen atoms; (d) indicates the axis system referred to in the text.

Details of the energy level pattern in a particular case depend, as in the case of a symmetric top, on such quantities as barrier height, separation of minima and the reduced mass for the vibration. Determination of this latter quantity, however, is usually more complicated in the case of an asymmetric rotor.

The reduced mass[9] depends on the path followed by the nuclei during the vibration and limiting cases for this may be difficult to determine in a complex case. The molecule NH_2X is a fairly typical example of an inverting asymmetric rotor and some possible classical paths for inversion are shown in Fig. 7.6. The reduced mass corresponding to each path may be determined through the definition $2T = \mu v^2$ where v is the magnitude of the velocity vector of a hydrogen atom in a molecule fixed axis system such as that shown in Fig. 7.6. Expressions for μ for the selected classical paths may be written in terms of geometric parameters. These expressions are lengthy and will not be given here, but essentially they are generalizations of equations (7.2) for the symmetric rotor case.

7.3.1 Rotation–inversion spectrum of an inverting asymmetric top molecule

The pattern of rotational energy levels associated with each vibrational state of an asymmetric top molecule has already been described in Section 2.8.3. In addition to the more complicated energy level scheme, the selection rules for an asymmetric rotor admit the possibility of transitions occurring which are not allowed in the symmetric rotor case, therefore it will be appreciated that spectra of inverting asymmetric top molecules can become quite complex.

The selection rules for transitions between the inversion states of the molecule depend on the transition moment integral,

$$S = \int \psi_v^* \mu_g \psi_{v'} \, d\tau \qquad (7.9)$$

where μ_g is the component of the molecular dipole moment along the gth principal axis of the molecule. When μ_a is antisymmetric with respect to inversion then ψ_v^* and $\psi_{v'}$ must be of opposite parity for the integral (7.9) to be non-zero. In the case of an asymmetric rotor, however, there may be other non-zero components of the dipole moment which will be symmetric with respect to inversion and in such a case ψ_v^* and $\psi_{v'}$ in equation (7.9) must have the same parity for a transition to be allowed. The selection rules for transitions between rotational states of the molecules are $\Delta J = 0, \pm 1$

Table 7.3   Selection rules for K_{-1} and K_{+1} for an asymmetric top molecule

Dipole component	Selection rules[a]			
	K_{-1}	$K_{+1} \longleftrightarrow K'_{-1}$		K'_{+1}
μ_a	$\begin{cases} e \\ 0 \end{cases}$	$\begin{matrix} e \\ e \end{matrix}$	$\begin{matrix} e \\ 0 \end{matrix}$	$\begin{matrix} 0 \\ 0 \end{matrix}$
μ_b	$\begin{cases} e \\ 0 \end{cases}$	$\begin{matrix} e \\ e \end{matrix}$	$\begin{matrix} 0 \\ e \end{matrix}$	$\begin{matrix} 0 \\ 0 \end{matrix}$
μ_c	$\begin{cases} e \\ e \end{cases}$	$\begin{matrix} e \\ 0 \end{matrix}$	$\begin{matrix} 0 \\ 0 \end{matrix}$	$\begin{matrix} e \\ 0 \end{matrix}$

[a] e and 0 refer to whether K_{-1} and K_{+1} are even or odd.

and the permitted changes in the pseudo-quantum numbers K_{-1} and K_{+1} are summarized in Table 7.3.

The consequences of these selection rules in terms of the allowed transitions for a typical prolate, inverting asymmetric top molecule (e.g. NH_2—CN) are illustrated in Fig. 7.7. In this example the μ_a component of the dipole moment is symmetric with respect to inversion and so rotational transitions involving this dipole component connect rotational states within the same inversion state. The μ_c component is, on the other hand, antisymmetric and rotational transitions involving μ_c occur between rotational states in inversion states of opposite parity. In this particular case $\mu_b = 0$ because of the symmetry of the molecule. However, in general μ_b too may be finite and give rise to additional possible transitions.

It will be evident from Fig. 7.7 that the details of the rotation–inversion spectrum for a molecule of this type depend very much on the magnitude of the inversion splitting (Δ) and thus on the barrier height. Some of the transitions between rotational states in a given inversion level will usually fall in the microwave region of the spectrum, however, the frequencies of the rotation–inversion transitions depend directly on Δ. When a molecule has a very high barrier to inversion, Δ tends to zero and the frequencies of the rotation–inversion transitions approach the corresponding frequencies of the pure rotational transitions associated with the equivalent dipole component. Consequently some of these transitions may appear as pure rotational transitions in the microwave region. When the inversion barrier becomes very low, Δ becomes large and most of the rotation–inversion transitions are

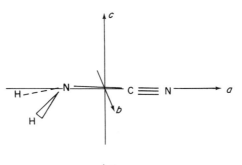

(a)

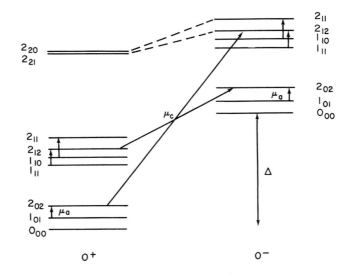

(b)

Fig. 7.7   (a) Principle axis system of cyanamide. (b) Rotational and rotation—inversion transitions in the 0^+ and 0^- states of cyanamide. The energy level manifolds are a diagrammatic representation (not to scale) of the situation in ND_2CN with $\Delta \sim 15$ cm^{-1}. The arrows within a given state show typical μ_a transitions while those from $0^- \leftarrow 0^+$ show two possible μ_c transitions. The dashed lines connect levels which strongly perturb each other (see text). Connections to other levels which will perturb more weakly have been omitted.

shifted out of the microwave region of the spectrum into the infrared. An interesting case arises when the barrier to inversion is moderately high and the magnitude of Δ becomes comparable with the splittings between the rotational levels of the molecule. Under these circumstances several of the rotation–inversion transitions may appear in the microwave spectrum as doublets separated by approximately 2Δ (see for example the $2_{12} \leftarrow 2_{02}$ μ_c transition of Fig. 7.7). In such cases identification of these transitions in the microwave spectrum of the molecule leads to a direct estimate of the splitting of the lowest inversion levels. In cases where the rotation–inversion transitions cannot be observed or have not been identified, an estimate of Δ can often be obtained from the pure rotational transitions themselves. When Δ is of the order of 50–100 cm^{-1} the upper inversion state will be considerably populated and so, since the rotational constants will not differ greatly between the 0^+ and 0^- states in these circumstances, each pure rotational transition will be accompanied by an intense vibrational satellite. The energy separation between the inversion states can be estimated from the relative intensities of the ground and excited state transitions, taking due account of any relevant statistical weight factors which may affect the population of the states. As described in Chapter 3, an estimate of the energy difference obtained by this method is unlikely to be as precise as that which can be obtained from frequency measurements of rotation–inversion transitions.

In order to determine a potential function governing inversion in a given molecule it is necessary to obtain experimental data in addition to the inversion frequency of the molecule itself. Often the inversion frequency of a suitable isotopic modification of the molecule can be obtained, however in order to define the potential function as well as possible it is preferable to obtain some information concerning the higher inversion states, e.g. 1^+ and 1^-. This data, when it is available, often comes from the infrared and far-infrared spectrum of the molecule. The potential function is then chosen to fit the data as closely as possible.

In practice, however, potential functions for the inversion vibration in amines have often to be determined using only data for the $0^{\pm}$ states owing to the complete absence of data for higher inversion states. Unfortunately this results in considerable uncertainties in the derived potential functions since this data gives only the area under the barrier above the energy levels.

7.3.2 Vibration–rotation interactions

It can be seen from Fig. 7.7 that provided the inversion frequency (Δ) is of the same order of magnitude as the spacing between rotational states in a molecule, accidental near degeneracies may occur between rotational energy levels in different vibrational states. Subject to limitations imposed by symmetry, nearly degenerate rotational levels may perturb each other through a vibration–rotation interaction similar in origin to the Coriolis interactions which occur for vibrational states (Section 2.9). In cases where perturbations of this kind are significant, the rotational energy level pattern in both vibrational states may suffer considerable modification which may manifest itself in deviations from rigid rotor behaviour of rotational transitions. It has already been pointed out that in these circumstances the rotational energy level manifolds in the different vibrational states of the molecule can no longer be treated as being independent and energy level calculations become correspondingly more complex.

As an illustration we will consider inversion in a molecule in which there is only one large amplitude mode generating internal angular momentum in the direction of only one of the principal axes. An example of such a molecule is cyanamide, see Fig. 7.7, where the internal angular momentum is parallel to the b axis. If it is assumed that any interaction between the inversion mode and other molecular vibrations can be neglected then the Hamiltonian for the problem has the form of equation (2.61),[10]

$$H = A_v P_a^2 + B_v P_b^2 + C_v P_c^2 + (\tfrac{1}{2} F p_v^2 + V(q)) - G p_v P_g \qquad (7.10)$$

The term in brackets in expression (7.10) represents a purely vibrational problem which can be solved using suitable wave functions, ϕ_v, and one of the potential functions described in Section 7.2.2 for the inversion. Usually any dependence of the reduced mass on the coordinate q for the vibration which appears in the coefficient F is small enough to be neglected where hydrogen atoms are mainly involved.

The Hamiltonian (7.10) implies that the axis system used in its derivation corresponds to the instantaneous principal axis system of the molecule and that the internal angular momentum appears entirely in the last term which represents the coupling with the component P_g of the overall angular momentum. Other axis systems may be chosen which lead to slightly different, although equivalent forms for the Hamiltonian.

The energy matrix associated with the remainder of expression (7.10) after removal of the vibrational part, can be conveniently set up using wave functions of the form,

$$\Psi = \psi(JKM)\phi_v$$

where $\psi (JKM)$ are asymmetric rotor wave functions. For the one axis coupling case the energy matrix has its simplest form if the z-axis of the limiting symmetric rotor is chosen to coincide with the axis g along which the internal angular momentum is orientated. In cyanamide this choice would correspond to $z = b, x = c, y = a$. As described in Section 2.9 the energy matrix contains terms connecting rotational states in inversion states of opposite parity which are introduced by the last term in the Hamiltonian. Usually simplification of the matrix can be effected by using a Van Vleck transformation to reduce to higher order matrix elements which connect nearly degenerate inversion levels with higher states. In this way the $0^{\pm}$ block for example may be uncoupled from the remainder of the energy matrix and treated independently. When this has been done the energy matrix for the 0^{+} and 0^{-} inversion states will have the form described in Section 2.9 and, for a molecule like cyanamide with $g = b$ and the choice of axes indicated above, the matrix for $J = 1$, for example, is identical with expression (2.63). Since the symmetry of P_g restricts the finite matrix elements introduced by the coupling term in the Hamiltonian to those connecting asymmetric rotor states of opposite parity,[11] the states in the 0^{+} and 0^{-} levels of cyanamide which can suffer perturbation as a result of the vibration–rotation interaction are typified by those connected by the dashed lines in Fig. 7.7.

Having derived the form of the energy matrix for any particular molecule the variables A_v, B_v, C_v, Δ and G which define the energy levels can be varied until the calculated rotation-vibration spectrum matches the observed one. This procedure gives a value for the inversion frequency which may subsequently be used together with other data to calculate a potential function for the vibration. An important corollary to these remarks is that even in cases where there is no obvious departure from a rigid rotor spectrum the vibration–rotation interactions may contribute significantly to the effective rotational constants A_v, B_v and C_v for a molecule and so caution must be exercised when deducing structural data from rotational constants obtained in these circumstances.

7.3.3 Inverting molecules

Effects attributable to inversion have been observed for many molecules. Amongst the best known examples are the substituted ammonias of the forms XNH_2 or X_2NH, and some parameters associated with potential functions for a few of the more thoroughly studied species are summarized in Table 7.4.

Aniline is an interesting example of an inverting molecule. For many years the question of the planarity of this molecule was an open one, however,

Table 7.4   Parameters relating to amine inversion in various molecules

Molecule	ϕ^a/deg	Barrier/kJ mol^{-1}	Type of potential function
NH_3	62	24·29	HO + Gaussian
		24·82	Manning
NH_2Cl	66	>42–48	HO + Gaussian
			Manning
NH_2NO_2	52	11·36	Manning
$NH_2C_6H_6$	37 (m.w.)		
	46 (u.v.)	6.76	HO + Gaussian
NH_2CN	38	5·58	Manning
NH_2CHO	0	0	Quadratic + Quartic

^{a} ϕ is the angle between the bisector of the HNH angle and the extension of the N—X bond.

microwave and ultraviolet spectroscopic evidence shows clearly that the X—NH_2 fragment of the molecule has a pyramidal configuration around the nitrogen atom. The microwave spectra of some thirteen isotopic species of aniline have been studied, allowing the determination of a very accurate structure and all the data have been found to be consistent with a pyramidal NH_2 fragment with the bisector of the HNH angle making an angle of 37° with the plane of the benzene ring. Relative intensity measurement of rotational transitions in the 0^+ and 0^- inversion states indicate that the inversion frequency for aniline itself is close to 46 cm^{-1}. These conclusions are supported by a vibrational analysis of the 294 nm band of aniline vapour. Bands associated with the inversion vibration in the ground and first excited electronic states of normal and several deuterated anilines have been assigned, and these indicate that the inversion vibration may be less anharmonic in the excited electronic state than in the ground state. These data imply an out-of-plane angle of 46° for the NH_2 fragment in the ground electronic state and the experimentally determined inversion splittings of aniline NH_2 and ND_2 have been used to determine a harmonic oscillator–Gaussian barrier potential function, with a barrier to inversion close to 6·8 kJ mol^{-1} (Table 7.4).

Cyanamide (NH_2CN) has already been mentioned as an example of an inverting molecule. Effects due to vibration–rotation interactions of the type discussed in Section 7.3.2 have been observed in the rotational spectra of this molecule and its isotopic species. These effects, together with the relative intensities of rotational transitions in the 0^+ and 0^- inversion states indicate that the inversion frequency for NH_2CN is of the order of $50\ cm^{-1}$. The same parameter for NHDCN and ND_2CN has been shown to be $33\ cm^{-1}$ and $15\ cm^{-1}$ respectively. Infrared measurements give the $0^- \leftarrow 0^+, 1^+ \leftarrow 0^-$ and $1^- \leftarrow 0^+$ separations in NH_2CN as 44, 408 and $714\ cm^{-1}$ respectively, and when this data is fitted to a quadratic + quartic barrier function the barrier to inversion is found to be about $5.6\ kJ\ mol^{-1}$. Further evidence to support the use of this particular potential function for cyanamide is the fact that it predicts the inversion splitting in ND_2CN to be $14.7\ cm^{-1}$, in excellent agreement with experiment.

Another particularly interesting molecule is formamide. The microwave spectrum consists of a ground state spectrum accompanied by a single intense vibrational satellite spectrum. This was originally interpreted in terms of an anharmonic vibration associated with a double minimum potential function and the derived energy levels were fitted to a Manning potential with a barrier to inversion of $4.4\ kJ\ mol^{-1}$ and an out-of-plane angle for the amine group of $17°$. Recently, however, it has been shown that if a number of infrared transitions are included in the data and a potential function of the form of equation (7.6) is used then the resulting potential constants have a positive sign for the parameter a. As was pointed out in Section 2.6.2. this means that $V(q)$ is not a double minimum function but rather an anharmonic single minimum function. The dependence of the rotational constants on vibrational state is also well accounted for by this analysis supporting the conclusion that formamide is a planar molecule with a very anharmonic NH_2 wagging vibration.

7.4 Inversion in Molecules Containing more than One Large Amplitude Vibration

Some molecules, e.g. hydrazine[9] and methylamine,[12] have more than one group which is capable of executing a large amplitude vibration. In these circumstances the assumption, implicit in previous discussions, that the low frequency mode is separable from other vibrational modes of the molecule with little or no coupling between them, is no longer acceptable. Coupling between two or more low frequency vibrations can occur in these cases and must be included in any calculation of the energy levels. Such problems are therefore very complicated and for a detailed discussion the original papers should be consulted.

The case of hydrazine is an interesting one where the two amino groups may rotate around the N—N bond as well as invert (Fig. 7.8). This molecule has

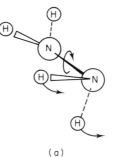

(a)

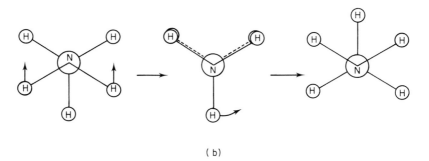

(b)

Fig. 7.8   Internal rotation and inversion in (a) hydrazine and (b) methylamine.

been studied by Raman, far-infrared and microwave spectroscopy and a theory for the complex internal motions has been derived. The barrier hindering the inversion vibration is around 2.8 kJ mol^{-1} whilst that hindering the internal rotational motion is near 3.1 kJ mol^{-1}. Methylamine presents a similar problem where an inversion of the amine group requires an internal rotation of the methyl group to return the molecule to a potential minimum as shown in Fig. 7.8. The barrier to internal rotation in this case is 8.18 kJ mol^{-1}.

7.5 Inversion in Excited Electronic States

Until now we have considered the possibility of the inversion vibration occurring only for molecules in their ground electronic state, however, there are several molecules which are known to be inverting in excited electronic states.

An example to which reference has already been made is aniline. In the first excited electronic state the reduced anharmonicity of the vibration is consistent with a reduction in the out-of-plane angle of the NH_2 fragment from

46° in the ground state to 30° in the excited state. Formaldehyde is in the position of being planar in the ground electronic state but non-planar in the first excited electronic state. An analysis of the vibrational levels in the excited state reveals that the relative energies of the lowest vibrational states associated with the out-of-plane deformation are 0, 127, 530 and 951 cm^{-1}. This highly anharmonic pattern has been interpreted to mean that the molecule is essentially non-planar in the excited state, the configuration around the carbon atom being pyramidal with an out-of-plane angle of about 20°. The double minimum potential associated with the CH_2 wagging motion has a barrier to inversion of around 1·9 kJ mol^{-1}.

7.6 Molecules with High Barriers to Inversion

The discussion in the foregoing sections has been mainly concerned with inversion in molecules where the barrier hindering the motion is less than about 25 kJ mol^{-1}. However, many molecules are known where the barrier is considerably higher than this.[13] For example, replacement of the hydrogen atoms of ammonia and many amines with suitable substituents can have the effect of raising the barrier and stabilizing the pyramidal configuration. Nuclear

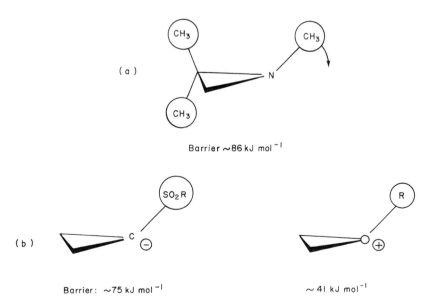

(a)

Barrier ∼86 kJ mol^{-1}

(b)

Barrier: ∼75 kJ mol^{-1} ∼ 41 kJ mol^{-1}

Fig. 7.9   (a) Inversion in aziridene rings is usually associated with a high barrier. (b) Inversion at carbanion and oxonium centres.

magnetic resonance spectroscopy has proved particularly useful in studying inversion in these high barrier cases using the procedures outlined in Chapter 3.

Incorporation of the nitrogen atom in a small ring, as in the aziridines for example, raises the inversion barrier as shown in Fig. 7.9(a). The increase in barrier height is attributed to the fact that the planar configuration of such a molecule would represent an extremely strained configuration and would therefore make inversion via the planar intermediate an energetically un-favourable process. Replacement of hydrogen atoms in ammonia by electron attracting substituents (e.g. CH_3 group) seems to reduce the barrier, while the effect of electronegative substituents such as halogens, oxygen, etc., around the nitrogen is usually to stabilize the pyramidal configuration of the molecule and raise the barrier.

Inversion at atoms other than nitrogen has also been studied although not to the same extent. Barrier heights for inversion at carbanion and oxonium centres are also given in Fig. 7.9(b) for comparison with the nitrogen cases.

Further Reading

Both of the following texts are concerned with microwave spectroscopy of molecules and they include detailed accounts of the phenomenon of inversion.

1. J. E. Wollrab. "Rotational Spectra and Molecular Structure", Academic Press London and New York, 1967.
2. W. Gordy and R. L. Cook. "Microwave Molecular Spectra", Wiley, Chichester and New York, 1970.

References

1. E. B. Wilson Jr., J. C. Decius and P. C. Cross. "Molecular Vibrations", Chap. 3. McGraw-Hill, New York, 1955.
2. C. H. Townes and A. L. Schawlow. "Microwave Spectroscopy", Chap. 3. McGraw-Hill, New York, 1955.
3. C. J. H. Schutte. "The Wave Mechanics of Atoms, Molecules and Ions", Chap. 3, Arnold, London, 1968.
4. (a) H. C. Longuet-Higgins, *Mol. Physics*, **6**, 445 (1963).
   (b) J. K. G. Watson. *Canad. J. Physics*, **43**, 1996 (1965).
5. J. B. Coon, N. W. Naugle and R. D. McKenzie. *J. Mol. Spectroscopy*, **20**, 107 (1966).
6. S. I. Chan. D. Stelman and L. E. Thompson. *J. Chem. Phys.* **41**, 2828 (1964).
7. M. F. Manning. *J. Chem. Phys*, **3**, 136 (1935).
8. C. C. Costain. *Phys. Rev.* **82**, 108L (1951).
9. T. Kasayu. *Sci. Papers Inst. Phys. Chem. Research (Tokyo)*, **56**, 1 (1962).
10. D. R. Lide Jr. *J. Mol. Spectroscopy*, **8**, 142 (1962).
11. V. Dobyns. *J. Chem. Phys.* **43**, 4534 (1965).
12. D. Kivelson and D. R. Lide Jr. *J. Chem. Phys.* **27**, 353 (1957).
13. H. Kessler. *Angew. Chem.* (Internat. Edit.), **9**, 219 (1970).

8
Large Amplitude Vibrations in Ring Compounds

8.1 Introduction

In previous chapters we have been concerned with motions in which one part of the molecule moves relative to the rest. It has been possible to discuss internal rotation to a high degree of approximation in terms of a rigid top and frame. In cyclic molecules constraints are imposed on the vibrations of the ring atoms and this leads to modes of vibration involving the entire skeleton of the molecule. For example, during the puckering vibration of the planar four-membered ring shown in Fig. 8.1 the internal angles of the ring become narrower

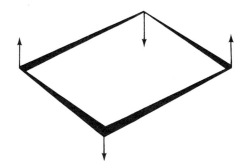

Fig. 8.1   The puckering vibration of a four-membered ring.

and twisting about the ring bonds also occurs. If a cyclic molecule of this type has a non-planar equilibrium conformation then the puckering vibration results in ring inversion (Fig. 8.2).

A planar or near planar ring consisting of n atoms has $n - 3$ out-of-plane vibrations. In the case of a five-membered ring these vibrations lead to a bend-

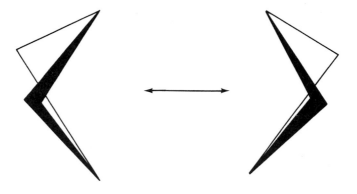

Fig. 8.2   The inversion of a four-membered ring.

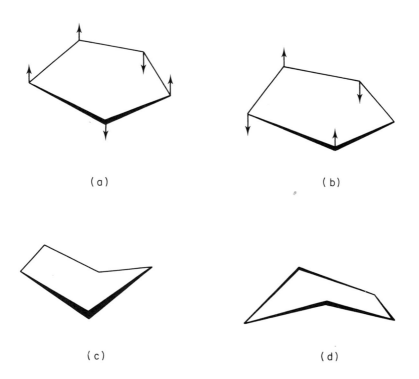

(a)   (b)

(c)   (d)

Fig. 8.3   The bending and twisting vibrations (a) and (b) and the bent (c) and twisted (d) conformations of a five-membered ring.

ing (Fig. 8.3(a)) and a twisting (Fig. 8.3(b)) of the ring. If the potential function associated with the out-of-plane motion of the ring atoms is asymmetric or a multi-minimum function the ring does not have a planar equilibrium conformation and the equilibrium conformations are referred to as the bent or envelope form (Fig. 8.3(c)) and the twisted or half chair form (Fig. 8.3(d)) respectively. As rings get larger and substituents are introduced the number of possible conformations increases rapidly. The study of the stability and chemical reactivity of ring compounds is one of the most important aspects of conformational analysis in organic chemistry.

The first experimental evidence that the ring puckering vibrations in cyclobutane[1] and cyclopentane[2] were unusual motions came from entropy measurements. Very detailed studies of polar derivatives of these two molecules have now been made using microwave and far-infrared spectroscopy and it is from these two techniques that most of the present information about puckering and inversion in four- and five-membered rings has been derived. The potential functions for cyclobutane and cyclopentane have been obtained from measurements on combination bands in the mid-infrared and from the Raman spectra of the vapours and the results confirm the conclusions drawn from thermodynamic work. Electron diffraction experiments have provided accurate molecular structures for a number of four- and five-membered ring molecules and again confirm the nature of the ring puckering vibrations.

Barriers to ring inversion in cyclohexane and related molecules fall in the range 30 to 40 kJ mol⁻¹ and a large number of barriers has been obtained using NMR and ultrasonic methods.[3] Microwave, electron diffraction and vibrational spectroscopic studies have been made on this type of molecule but the information derived is not as detailed as that for four- and five-membered rings.

Electron diffraction studies of a number of seven-membered and larger rings have shown some of these molecules to be remarkably flexible. X-ray crystallography has also been used to establish the most stable conformations of crystalline samples of medium and large ring compounds, but it must be remembered that these are not necessarily the most stable forms in the liquid state or in solution.

In this chapter the question of how ring molecules pass from one conformation to another and how the potential functions for these motions may be obtained is discussed. Most attention will be given to small ring compounds because of their relative simplicity. The conclusions which may be drawn about such factors as strain in the ring bond angles, torsional strain and transannular interactions are of interest because of their possible extrapolation to large molecules.

8.2 Strain in Ring Compounds

The conformations and the nature of the skeletal vibrations of cyclic molecules are often the result of a delicate balance between a number of different types of forces. In conformational analysis three different sources of ring strain are usually distinguished:

(a) angle or Bayer strain
(b) torsional or Pitzer strain
(c) transannular interactions.

The first type stems from the explanation given by von Bayer in 1885 for the relative instability of three- and four-membered rings compared to five- and six-membered rings. He argued that considerable energy must be required to compress the CCC bond angles from the tetrahedral values found in acyclic molecules to give internal angles of 60° in cyclopropane and 90° in the planar ring form of cyclobutane. Although modern valence theory provides a description of these molecules in terms of bent or "banana" bonds, angle strain still plays an important part in the interpretation of potential functions for ring puckering vibrations. If a planar ring is bent the internal angles decrease and a measure of the increase in the strain of the ring may be obtained by associating force constants with the changes in the internal angles.

Torsional strain arises from non-bonded interactions between substituents attached to adjacent ring atoms and is, as its name implies, connected with twisting about bonds within the ring. It is called Pitzer strain because of the pioneering work of K. S. Pitzer on internal rotation about carbon–carbon single bonds. Figure 8.4. shows perspective views along a carbon–carbon bond in the planar and non-planar ring conformations of a molecule such as cyclobutane. In Fig. 8.4(a) the situation is similar to that in the eclipsed form of ethane and represents a high torsional energy, while Fig. 8.4(b) is closer to the staggered form of ethane and has a lower torsional energy. By associating a simple cosine potential with the twisting about the ring bonds it is possible to estimate the changes in the torsional strain as the ring is distorted.

Transannular interactions occur between substituents attached to ring atoms that are not directly connected by a bond. Interactions of this type are responsible for the stability of equatorial monosubstituted cyclohexanes compared to the corresponding axial conformers. They are also important in accounting for the relative thermochemical instability of nine, ten- and eleven-membered rings compared to cyclohexane and cycloduodecane.

The concept that potential functions for skeletal vibrations are the result

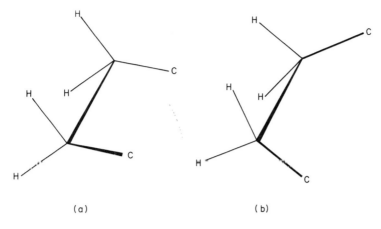

(a)   (b)

Fig. 8.4   Perspective views along a carbon–carbon bond in (a) the planar (b) a non-planar ring conformation of cyclobutane.

of a balance between competing forces may be used to show how single and double minimum functions may occur.[4] Figure 8.5(a) shows the in-phase addition of a single minimum curve (dotted line) representing the angle strain and a periodic curve (full line) representing the torsional strain to produce a single minimum function. The out-of-phase addition of these curves, shown in Fig. 8.5(b), produces a double minimum function.

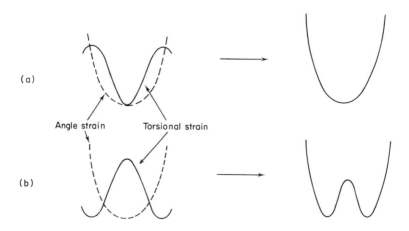

(a)

Angle strain   Torsional strain

(b)

Fig. 8.5   Showing how angle and torsional strains may combine to produce (a) a single minimum and (b) a double minimum potential function.

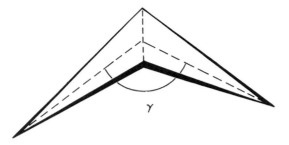

Fig. 8.6   The dihedral angle for the puckering of a four-membered ring.

The ideas expressed in the previous paragraph may be put into more quantitative form in order to estimate whether a molecule such as cyclobutane is likely to have a planar or non-planar ring equilibrium conformation. If the ring angle bending potentials are assumed to be harmonic then the angle strain will be given by a sum of terms of the type $\frac{1}{2}K_\theta(\Delta\theta)^2$ where K_θ is an angle bending force constant and $\Delta\theta$ is the change in the ring angle that occurs as a result of the puckering vibration. The torsional strain may be written as a sum of terms of the type $V_3/2(1 - \cos 3\alpha)$. The total potential energy may be

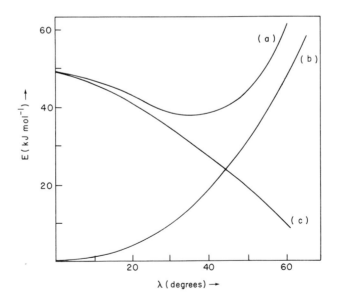

Fig. 8.7   The variation of (a) the potential energy (b) the angle strain and (c) the torsional strain for the puckering of cyclobutane. ($\lambda = 180° - \gamma$).

expressed as,

$$V = \frac{1}{2} \sum_{\substack{\text{ring} \\ \text{angles}}} K_\theta (\Delta\theta)^2 + \frac{1}{2} \sum_{\substack{\text{ring} \\ \text{bonds}}} V_3 (1 - \cos 3\alpha) \tag{8.1}$$

The degree of puckering of the ring may be conveniently measured using the dihedral angle γ shown in Fig. 8.6. The contributions from angle strain and torsional strain and the total ring strain energies for cyclobutane are shown in Fig. 8.7. While calculations of this type are not necessarily very accurate they do show that barriers to inversion in four-membered rings are unlikely to be more than a few $kJ\,mol^{-1}$.

8.3 Four-Membered Rings

8.3.1 Coordinates and potential functions for four-membered rings

The potential function for the puckering of a four-membered ring can be expressed in terms of the dihedral angle γ shown in Fig. 8.6, however, it is generally more convenient to use a distance rather than an angle as a variable parameter. The coordinate, x, shown in Fig. 8.8, defined as one half of the separation of the ring diagonals measured through the point where they cross when the ring is planar, has been widely used. The question now arises of whether the changes in potential energy that arise from angle, torsional and other sources of ring strain may be expressed as a simple function of x.

During the 1940's there was considerable interest in diborane and a cyclic structure with bridging hydrogen atoms (Fig. 8.9) was proposed for this molecule, and it was in connection with these studies that the first potential function for a ring puckering vibration was proposed.[5] It was pointed out that for small deviations from planarity the square of the ring puckering coordinate is proportional to the change in the internal angle of the ring (θ).

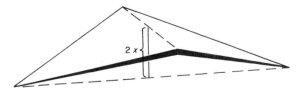

Fig. 8.8 The ring puckering coordinate for a four-membered ring.

Fig. 8.9   The cyclic structure of diborane.

If the ring puckering potential energy arises solely from the harmonic bending of the ring angles then,

$$V(x) = ax^4$$

The idea of a quartic potential has played an important part in the study of four-membered rings.

The simplest potential which is predominantly quartic but allows the possibility of a double minimum function is,

$$V(x) = ax^4 - bx^2 \qquad (8.2)$$

When b is negative this potential has a single minimum. When b is positive there are two minima at $x = \pm(b/2a)^{\frac{1}{2}}$ and the height of the central barrier is $b^2/4a$. The two parameter function, equation (8.2), has been found to reproduce the observed ring puckering energy levels of a number of four-membered rings very accurately and indeed it is thought to represent a more useful function than other types of expressions containing three or more adjustable parameters.

Monosubstituted cyclobutanes have the possibility of existing as non-equivalent equatorial and axial conformers (Fig. 8.10) and consequently equation (8.2) must be modified to allow for the resultant asymmetry in the potential function. For a number of molecules this has been done by adding a cubic term to equation (8.2).

$$V(x) = ax^4 - bx^2 + cx^3 \qquad (8.3)$$

The origin of $V(x)$ occurs at the central maximum where $dV/dx = 0$, and this does not necessarily coincide with the planar ring conformation. A linear term may be included in equation (8.3) to change the origin, but usually this is neglected and the origin of the curve is assumed to occur at the planar conformation.

The Hamiltonian operator for the ring puckering vibration may be written

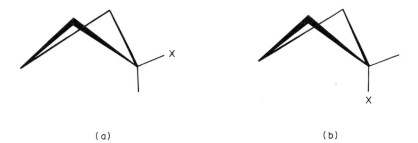

(a)   (b)

Fig. 8.10   (a) The equatorial and (b) axial conformations of a monosubstituted cyclobutane.

in terms of the coordinate x as,

$$H = -\frac{\hbar^2}{2\mu}\frac{\mathrm{d}^2}{\mathrm{d}x^2} + V(x) \tag{8.4}$$

where μ is the reduced mass for the vibration and $V(x)$ is given by either equation (8.2) or (8.3). The energy levels and wave functions of equation (8.4) are usually calculated by setting up and diagonalizing the Hamiltonian matrix using a convenient set of basis functions as described in Chapter 2. The wave functions of both the simple harmonic oscillator and the quartic oscillator have been used for this purpose. The calculation of the reduced mass presents some problems since it depends both on the coordinate x and the paths along which the vibrating atoms move. This has been investigated in for example oxetane and thietane[6] through analysis of the vibration–rotation interaction. Consideration of a number of possible motions for both molecules in which the bond lengths do not change shows that the closest agreement with the experimental results is obtained for a motion which is essentially a bending about the ring diagonals.

The close balance of the various types of competing intramolecular forces in four-membered rings is extremely sensitive to the extramolecular environment. Considerable changes have been observed in the Raman spectrum of cyclobutane in going from solid to liquid to gas.[7] It is therefore important to evaluate potential functions from data obtained from the gas phase, especially if potential functions for different molecules are to be compared.

The frequency of the ring puckering fundamental vibration of cyclobutane is smaller by a factor of nearly four than that of the next lowest vibration of the same symmetry. The separation of the ring puckering vibration and its treatment as a one-dimensional problem is therefore a reasonable approximation.

However, a small, but real difference found for the potential functions of cyclobutane and cyclobutane-d_8 implies that the one parameter potential is not strictly correct. The separation is expected to be a poorer approximation in substituted cyclobutanes containing a number of heavy atoms because the external angle bending and rocking vibrations will have much lower frequencies than in cyclobutane.

8.3.2 Examples of ring puckering

Four different examples of ring puckering potential functions are shown in Fig. 8.11. These represent examples of,

(a) a symmetrical single minimum,
(b) a symmetrical double minimum,
(c) an asymmetrical single minimum,
(d) an asymmetrical double minimum.

The figure also includes the observed energy levels which have been used to derive these functions. Table 8.1 summarizes the experimental data on the puckering vibration for a number of four-membered rings with symmetrical and unsymmetrical potential functions.

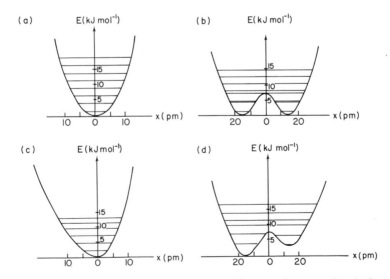

Fig. 8.11   The ring puckering potential functions and energy levels for (a) a symmetrical single minimum, e.g. oxetanone, (b) a symmetrical double minimum, e.g. cyclobutane, (c) an unsymmetrical single minimum, e.g. bromocyclobutane, and (d) an unsymmetrical double minimum, e.g. azetidine.

Table 8.1 Equilibrium conformations and barriers to inversion of some four-membered rings with symmetric and unsymmetric potential functions.

X	Y		x_{min}/pm[a]	Barrier/kJ mol^{-1}
CH_2	CH_2	cyclobutane	17	6·11
CH_2	O	oxetane	7	0·17
CH_2	S	thietane	15	3·26
CH_2	Se	selenetane	17	4·48
CH_2	SiH_2	silacyclobutane	18	5·27
CH_2	C=O	cyclobutanone	6	0·04
O	C=O	oxetanone	0	—
S	C=O	thietanone	0	—
O	$C=CH_2$	methylene oxetane	0	—
CH_2	NH	azetidine	$\begin{cases} +14 \\ -15 \end{cases}$	$\begin{cases} 5·27 \\ 4·14 \end{cases}$
CH_2	CHCl	chlorocyclobutane	~22	—
CH_2	CHCN	cyanocyclobutane	~0	—

[a] x is defined in Fig. 8.8.

The results in Table 8.1 can be rationalized to a large degree using the concept of competing angle and torsional strains. Cyclobutane and silacyclobutane are non-planar in their equilibrium conformations, and with four pairs of eclipsing hydrogen atoms in the planar ring conformation, have the highest barriers to ring inversion. The molecules with three methylene groups also have non-planar ring equilibrium conformations. However, the sequence of barriers to ring inversion in oxetane, thietane and selenetane does not follow the observed trend in barriers to internal rotation for methanol ($V_3 = 4·5$ kJ mol^{-1}), methanethiol ($V_3 = 5·3$ kJ mol^{-1}) and methyl selenol ($V_3 = 4·2$ kJ mol^{-1}).

Cyclobutanone represents an example where an extremely fine balance of the angle and torsional strains occurs. Because of the sp^2 hybridization of the carbon atom of the carbonyl group, angle strain will increase more rapidly with ring puckering than in the other molecules. This factor is counterbalanced by the torsional strain of three eclipsing methylene groups and the result is a vanishingly small barrier to ring inversion. Both oxetane and cyclobutanone pose the largely philosophical question of whether the ring is planar since for both molecules the barrier to ring inversion is less than the zero point vibrational energy.

There has been a considerable debate in the literature over the existence of more than one conformer for the monosubstituted cyclobutanes.[8,9] Infrared spectroscopic evidence suggests the existence of two conformers but in each case the pure rotational spectrum of the equatorial rotamer only has been detected by microwave spectroscopy. The far-infrared spectra of a number of these molecules have now been observed and ring puckering transitions assigned.[10] The potential functions derived from this data have an asymmetric single minimum and leave little doubt that only the equatorial conformation exists. Azetidine is the first example of a four-membered ring compound to show an unsymmetrical double minimum potential function.

Ring puckering transitions have also been observed in the far-infrared and Raman spectra of diborane and an analysis of the transition frequencies give a potential function which is dominated by the quadratic term, in contradiction with the earlier hypothesis that this molecule would have a predominantly quartic potential.

8.4 Five-Membered Rings

Five-membered rings have two skeletal vibrations and these may interact with each other to produce vibrational motions which are very much more complicated than those that have been discussed so far. In order to understand how these may arise it is necessary to consider the possible two dimensional potential functions which may be associated with the ring bending and twisting vibrations. The deviation of the ring from planarity may be expressed in terms of two orthogonal coordinates x_b and x_t which represent the degrees of bending and twisting of the ring. For the present we will not enquire further about the physical nature of these coordinates since there is a certain arbitrariness in choosing them.

The potential functions may be classified in the following way depending on whether they have single or double minima with respect to the coordinates x_b and x_t:

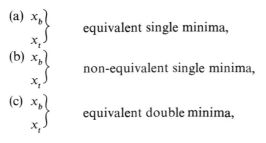

(a) $\left.\begin{array}{l} x_b \\ \\ x_t \end{array}\right\}$ equivalent single minima,

(b) $\left.\begin{array}{l} x_b \\ \\ x_t \end{array}\right\}$ non-equivalent single minima,

(c) $\left.\begin{array}{l} x_b \\ \\ x_t \end{array}\right\}$ equivalent double minima,

(d) $\left.\begin{array}{c} x_b \\ \\ x_t \end{array}\right\}$ non-equivalent double minima,

(e) $x_b(x_t)$ double minima,

   $x_t(x_b)$ single minima.

The functions (a) and (b) do not give rise to any abnormalities in the energy level manifolds and therefore will not be discussed further. The planar conformation of cyclopentane would have a potential function of type (a), while cyclopentadiene has a function similar to (b).

The potential functions (c) and (d) may give rise to an unusual vibrational motion and associated pattern of energy levels. This is known as pseudorotation and was first invoked to explain the anomalous thermodynamic properties of cyclopentane. One of the important consequences of this type of motion is that it enables a ring molecule to invert without going through the high energy planar conformation. We shall see that a similar mechanism can be proposed for the inversion of cyclohexane and large ring molecules.

The potential function (e) is found in cyclopentene and a number of other five-membered rings which contain a double bond. In this type of molecule there is little interaction between the two skeletal vibrations and they may be treated separately to a good approximation.

8.4.1 Pseudorotation

Cyclopentane has ten equivalent bent conformations of C_s symmetry and ten equivalent twisted conformations of C_2 symmetry. Each is characterized by having a unique atom (Fig. 8.12), which in the case of a bent conformation lies in the C_s symmetry plane while for a twisted conformation it lies on the C_2 rotational axis. If the ring atoms are numbered in a cyclic manner then the ring conformations may be labelled by the symbol B or T, the number of the unique atom and the superscript $+$ or $-$ depending on whether the ring is normal or inverted with respect to a reference plane. The conformations shown in Fig. 8.12 would be referred to as $B1^+$ and $T1^+$ respectively.

For the bending and twisting of the ring shown in Fig. 8.13 the out-of-plane displacement coordinates for each atom of the ring may be expressed as,

$$\left.\begin{array}{l} x_{bk} = r\cos\left\{\dfrac{4\pi}{5}(k-1)\right\} \\[4mm] x_{tk} = -r\sin\left\{\dfrac{4\pi}{5}(k-1)\right\} \end{array}\right\} \quad k = 1, 2, 3, 4, 5$$

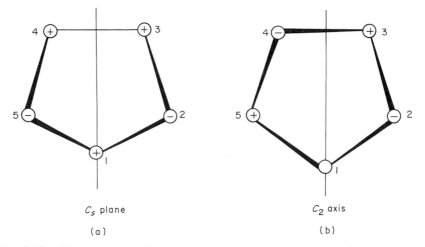

C_s plane C_2 axis

(a)   (b)

Fig. 8.12   The symmetry elements and the unique atoms of (a) bent and (b) twisted conformations of cyclopentane.

where x_{bk} refers to the displacement of the kth atom due to the bending mode and x_{tk} its displacement due to the twisting mode. The quantity r represents the maximum displacement of an atom in either the bending or twisting modes. If simultaneous bending and twisting of the ring takes place the total displacements may be written as an arbitrary linear combination of the two displacements x_{bk} and x_{tk} giving,

$$x_k = r \cos\left\{\frac{4\pi}{5}(k - 1) + \phi\right\}$$

where ϕ is a phase angle. By allowing ϕ to vary, the position of maximum out-of-plane displacement moves round the ring and this is shown for one complete cycle in Fig. 8.13. The molecule alternates between bent and twisted conformations at intervals of 18° and at $\phi = 180°$ the ring is inverted with respect to its original state. This has been achieved without ever passing through the planar conformation.

The reader will have noticed that the description has effectively changed from the coordinates x_b and x_t to the polar coordinates r and ϕ as defined in Fig. 8.14. It is interesting to discuss the nature of the motion of the molecule with respect to these new coordinates. For simplicity an isotropic two-dimensional double minimum potential function with a high central barrier will be assumed. This may be visualized by rotating Fig. 8.11(b) about the E axis. At $r = 0$, corresponding to the planar molecule, there is a high barrier;

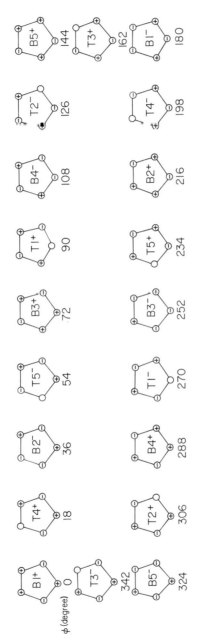

Fig. 8.13 The form of the molecular conformations during the course of one cycle of the phase angle ϕ for cyclopentane.

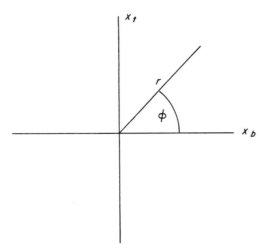

Fig. 8.14   The relationship between the coordinates x_b and x_t and the polar coordinates r and ϕ.

as r increases the potential energy reaches some minimum corresponding to the equilibrium puckered conformations of the ring and then increases again. The motion of the molecule with respect to the r coordinate is one in which the amplitude of puckering oscillates about the equilibrium puckered conformations of the ring, and is in all respects rather an ordinary vibration.

The motion with respect to the ϕ coordinate is very unusual since the potential energy is independent of this coordinate. This leads to a set of energy levels similar to those for free internal rotation. Since the actual displacements of the ring atoms x_k are perpendicular to the direction of movement of ϕ no angular momentum is generated by this motion. Because of these two factors the motion has been termed pseudorotation, and ϕ is often referred to as the pseudorotation coordinate.

The potential function mentioned in the previous paragraph implies that all the possible ring conformations for a fixed amplitude of the puckering coordinate (r) have the same energy. This is hardly likely to occur in a real molecule and allowance can be made for this by introducing periodic terms into the potential energy of the form $V_n/2\,(1 - \cos n\phi)$. The value of n depends on the number of equivalent conformations the molecule assumes during a cycle of the pseudorotation coordinate. In the case of cyclopentane if the bent and twisted conformers have slightly different energies the lowest value of n is 10, but if the two types of conformer do have exactly the same energy then the lowest value of n is 20. In derivatives of cyclopentane because the number of equivalent conformers is smaller values of n of 2 or 4 are common. The pseudorotational energy levels now become similar to those for hindered internal rotation and the motion is therefore referred to as hindered pseudorotation.

8.4.2 The Hamiltonian operator and energy levels for pseudorotation

In terms of the bending and twisting coordinates, x_b and x_t, the Hamiltonian operator for free pseudorotation may be written as,[11]

$$H = -\frac{\hbar^2}{2\mu}\left(\frac{\partial^2}{\partial x_b^2} + \frac{\partial^2}{\partial x_t^2}\right) + a(x_b^2 + x_t^2)^2 - b(x_b^2 + x_t^2) \qquad (8.5)$$

where the same reduced mass μ and potential function of the type given by equation (8.2) has been assumed for both vibrations. Using the relationships,

$$x_b = r \cos\phi$$
$$x_t = r \sin\phi$$

implied by Fig. 8.14 the operator may be transformed to give,

$$H = -\frac{\hbar^2}{2\mu}\left\{\frac{1}{r}\frac{\partial}{\partial r}\left(r\frac{\partial}{\partial r}\right) + \frac{1}{r^2}\frac{\partial^2}{\partial \phi^2}\right\} + ar^4 - br^2 \qquad (8.6)$$

When the central barrier in the potential function is high and the amplitude of the radial puckering about the equilibrium puckered conformation (r_{min}) is small equation (8.6) can be separated into radial and angular parts,

$$H_r = -\frac{\hbar^2}{2\mu}\left(\frac{1}{r}\frac{d}{dr} + \frac{d^2}{dr^2}\right) + V(r) \qquad (8.7)$$

$$H_\phi = -\frac{\hbar^2}{2\mu}\frac{1}{r_{min}^2}\frac{d^2}{d\phi^2} \qquad (8.8)$$

and the total energy is,

$$E = E_r + E_\phi$$

The potential $V(r)$ of equations (8.7) may be approximated by the harmonic expression,

$$V(r) = \tfrac{1}{2}k_r(r - r_{min})^2$$

and the energy levels of equation (8.7) are similar to those of a harmonic oscillator,

$$E_n = (n + \tfrac{1}{2})hv_r - \frac{\hbar^2}{8\mu r_{min}^2} \qquad n = 0, 1, 2, \ldots \qquad (8.9)$$

and n is referred to as the radial quantum number. The second term in equation (8.9) is part of the zero point energy. The energy levels of equation (8.8) are given by,

$$E_m = Bm^2 \qquad m = 0, \pm 1, \pm 2 \qquad (8.10)$$

and

$$B = \frac{\hbar^2}{2\mu r_{min}^2}$$

The constant B is usually called the pseudorotational constant and m the pseudorotational quantum number. The total energy is then,

$$E(n, m) = (n + \tfrac{1}{2})hv_r + B(m^2 - \tfrac{1}{4}) \qquad (8.11)$$

Stacks of vibrational–pseudorotational energy levels for the ground and first excited state of the radial mode are shown in Fig. 8.15(a). The energy levels for hindered pseudorotation may be calculated by introducing a potential consisting of one or several terms of the type $V_n/2(1 - \cos n\phi)$ into equation (8.8). The resulting energy levels for a relatively low twofold barrier are shown in Fig. 8.15(b). As in the case of hindered internal rotation the levels below the barrier are most highly perturbed.

The validity of the separation of the Hamiltonian operator into equations (8.7) and (8.8) has been tested by setting up the complete Hamiltonian using the wave functions of the isotropic two dimensional harmonic oscillator as a basis.[12] In the case of free pseudorotation, energy levels in the region of the central barrier are found to deviate considerably from those given by equation (8.11). For hindered pseudorotation the energy levels are found to be strongly dependent on both the height of the central barrier and the nature of the periodic potential.

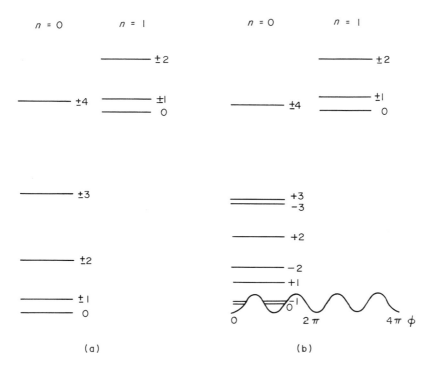

Fig. 8.15 The energy levels for (a) free pseudorotation (b) pseudorotation hindered by a low twofold barrier. The levels are labelled using the pseudorotational quantum number m.

A variety of transitions involving either or both of the radial and pseudorotational modes may be observed in the far-infrared, infrared, Raman and electronic spectra for molecules which exhibit pseudorotation. The types of transitions involved for any particular case depend on factors such as the symmetry of the molecule, the central barrier and the barrier to pseudorotation.

8.4.3 Examples of pseudorotation

Cyclopentane, tetrahydrofuran and 1,3-dioxolane have free or slightly hindered pseudorotation and a summary of the spectroscopic constants and potential functions for these molecules is given in Table 8.2. The degree of puckering in the equilibrium conformation is very similar in all three molecules and the trend in the barriers to planarity can be attributed to the reduction in torsional strain as methylene groups are replaced by oxygen

Table 8.2   Data pertaining to pseudorotation in some five-membered ring compounds.

	Cyclopentane	Tetrahydrofuran	1,3-Dioxolane
First radial transition/cm^{-1}	270	260	260
Pseudorotational constant/cm^{-1}	2·5	3·2	4·2
Equilibrium puckering amplitude/pm	48	44	38
Barrier to planarity/kJ mol^{-1}	22	13	8
Barrier to pseudorotation/kJ mol^{-1}	~0	0·69	0·13

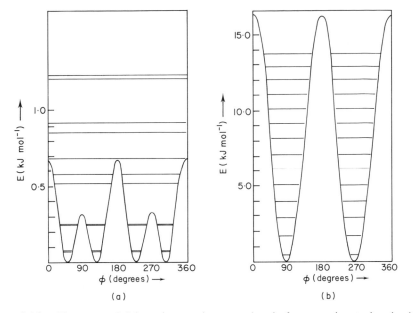

Fig. 8.16   The potential function and energy levels for pseudorotation in (a) tetrahydrofuran, and (b) silacyclopentane.

atoms. For cyclopentane, spectroscopic transitions involving the lowest pseudorotational levels have not been observed and so the possibility of a small barrier cannot be ruled out. A detailed study of the microwave spectrum of tetrahydrofuran has shown the potential function hindering pseudorotation to have the form illustrated in Fig. 8.16(a) with maxima at $T1^+$ and $T1^-$, $B1^+$ and $B1^-$ (Fig. 8.13) the former being lower in energy than the latter. The potential function for 1,3-dioxolane has a similar shape to that of tetrahydrofuran, however, in this instance the maxima in the potential energy curve occurring at the twisted configuration $T1^+$ and $T1^-$ are about 125 J mol^{-1} higher in energy than those at the $B1^+$ and $B1^-$ bent configurations.

Silacyclopentane and cyclopentanone are examples of molecules with high twofold barriers to pseudorotation with equilibrium conformations corresponding to the $T1^+$ and $T1^-$ forms of Fig. 8.13. The observed energy levels and potential function hindering the pseudorotation for silacyclopentane are shown in Fig. 8.16(b). Since all of the observed energy levels lie below the barrier, they are essentially doubly degenerate; in silacyclopentane the splitting of the lowest pair of levels is calculated to be 10^{-15} cm^{-1} while even the fourteenth excited state levels are predicted to be split by only 1 cm^{-1}. A number of other derivatives of cyclopentane of the type $(CH_2)_4X$, where X is a

H

symmetrical group, have been found to have twisted equilibrium conformations and fairly high barriers to pseudorotation.

For certain monosubstituted cyclopentanes there is evidence for the existence of both axial and equatorial conformations. This is the case for chlorocyclopentane and bromocyclopentane and these molecules also exhibit pseudorotation. Only radial mode transitions have been observed but these indicate that the barriers to pseudorotation are likely to be below $1\,kJ\,mol^{-1}$.

8.4.4 Five-membered rings with one or two double bonds

The presence of a double bond in cyclopentene makes it very much more difficult to twist the ring in this molecule compared to cyclopentane. This leads to a separation of the ring twisting and bending modes and consequently cyclopentene and related molecules do not show pseudorotation. There are many similarities between the out-of-plane bending vibration of a five-membered ring with a double bond and the puckering vibration of a four-membered ring and the former are often referred to as pseudo four-membered rings. The potential energy for the bending of a pseudo four-membered ring may be expressed in terms of the coordinate shown in Fig. 8.17 and the potential function is usually expressed in the form of equations (8.2) or (8.3).

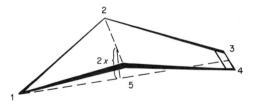

Fig. 8.17   The ring puckering coordinate for a five-membered ring containing one double bond.

The equilibrium conformations and, where appropriate, barriers to inversion for a number of five-membered ring molecules containing a double bond are given in Table 8.3. It can be seen that there is a greater tendency for pseudo four-membered rings to adopt a planar ring equilibrium conformation than the corresponding four-membered rings. This is in large part due to the fact that the torsional strain associated with the 2–3 and 4–5 bonds (Fig. 8–17) in the five-membered rings is minimized in the planar conformation.

Five-membered rings with two double bonds are generally very rigid molecules with planar ring equilibrium conformations. For example in a

Table 8.3   Barriers to inversion of some five-membered rings containing a double bond

X	Barrier/kJ mol^{-1}	Barrier/kJ mol^{-1}
CH_2	2·76	2·76
O	0·00	1·00
S	0·00	2·85
C=O	0·00	0·00

five-membered heterocyclic molecule such as imidazole the lowest frequency vibration of the ring atoms occurs at ~ 500 cm^{-1}. The rigidity of such species is partly due to electron delocalization over the ring.

8.5 Six-Membered and Larger Rings

8.5.1 Cyclohexane and its derivatives

The possible complications of having three interacting skeletal vibrations fortunately do not occur in the most stable conformations of cyclohexane and related molecules. The equilibrium conformations of these molecules are relatively free from angle and torsional strain and consequently the skeletal modes are near harmonic, small amplitude vibrations. The ring inversion of cyclohexane however is a more complicated process than for cyclobutane and cyclopentane and also the barrier is considerably higher, although, some similarities do exist between the inversion of five- and six-membered rings.

Cyclohexane and many of its simple derivatives have the chair equilibrium conformation (Fig. 8.18(a)) in which the angle and torsional strain are minimized. The boat conformation of cyclohexane (Fig. 8.18(b)) is also free of angle strain, but torsional strain, due to eclipsing methylene hydrogen atoms and transannular interactions between the "bowsprit" hydrogen atoms, make this conformation some 29 kJ mol^{-1} less stable than the chair form. The torsional and transannular strain in the boat form is alleviated to some extent in the twist-boat conformation (Fig. 8.18(c)), which has been estimated to be 7 kJ mol^{-1} more stable than the boat conformation.

It is generally agreed that the inversion of cyclohexane proceeds through a

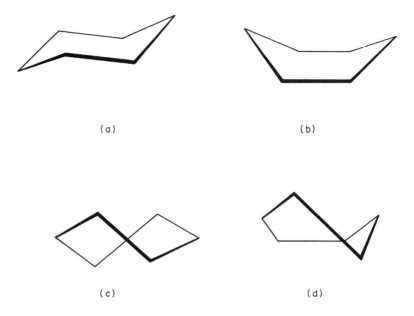

Fig. 8.18   (a) The chair (b) the boat (c) the flexible or twist boat and (d) the half chair conformations of cyclohexane.

transition state which corresponds to the "cyclohexene"-like half chair conformation (Fig. 8.18(d)) rather than the planar ring conformation. The former has been estimated to have an energy 53 kJ mol^{-1} above that of the chair conformation. The energetically most economical pathway for the inversion of cyclohexane has an associated potential energy function of the form shown in Fig. 8.19.

8.5.2 Six-membered rings with one and two double bonds

Cyclohexene and other six-membered rings with one double bond have the half chair equilibrium conformation shown in Fig. 8.18(d). The ring bending and twisting vibrations of 1,4-dioxene (Fig. 8.20(a)) and 2,3-dihydropyran (Fig. 8.20(b)) show some similarities to the corresponding vibrations of five-membered rings which undergo hindered pseudorotation. In both molecules these vibrations involve mainly movements of the methylene groups (Fig. 8.21). The third out-of-plane vibration is essentially a torsion about the double bond. It occurs at a higher frequency and does not interact with the other two skeletal modes. From the far-infrared and Raman spectra it has been possible to build up a detailed picture of the inversion of the half chair conformation in

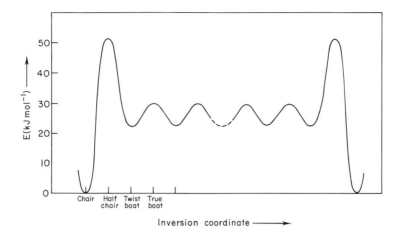

Fig. 8.19   An indication of the path of ring inversion in cyclohexane.

these molecules, and the lowest energy path for this process is shown in Fig. 8.22. As for cyclopentane and cyclohexane the molecules are able to invert without having to pass through the high energy planar ring conformation.

Six-membered rings with two double bonds have a great tendency to adopt a planar ring equilibrium conformation since the preferred valence angle for the four sp^2 hybridized ring atoms is $120°$. However, the balance of the competing forces is quite fine as illustrated by the fact that 1,3-cyclohexadiene has a non planar ring while 1,4-cyclohexadiene is planar. Molecules such as

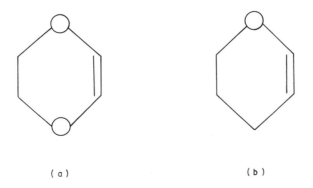

(a)   (b)

Fig. 8.20   The structure of (a) 1,4-dioxene and (b) 2,3-dihydropyran

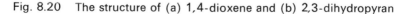

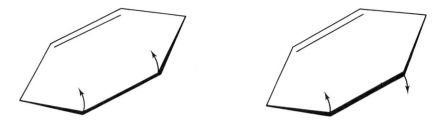

Fig. 8.21   (a) The bending and (b) twisting vibrations of a six-membered ring containing a double bond.

4-pyrone, where there are five sp^2 hybridized atoms in the ring, are planar, and benzene and its derivatives all have rigid planar rings.

8.6 Fused and Bridged Ring Systems

Large amplitude motions are not usually a feature of molecules containing fused or bridged ring systems. Exceptions are found when the presence of one ring does not interfere with the inversion of a second or where the whole molecule is forced to adopt a highly strained conformation. The nature of the

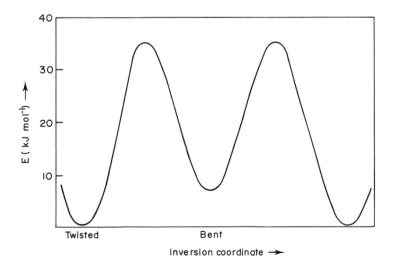

Fig. 8.22   The path of ring inversion in 1,4-dioxene.

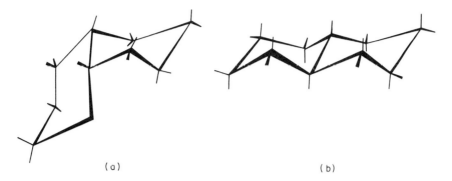

Fig. 8.23   The conformations of (a) *cis*-decalin (b) *trans*-decalin.

resulting motions will be illustrated by a brief discussion of particular examples.

The fact that decalin was shown to exist in *cis* and *trans* forms (Fig. 8.23) played an important part in convincing chemists that cyclohexane rings are puckered and not planar. Although both cyclohexane rings have chair conformations in these two forms of the molecule there is an interesting difference in their thermochemical stability and flexibility. In *trans*-decalin the two rings are linked through equatorial bonds while in *cis*-decalin they are linked through an axial bond and an equatorial bond. In *cis*-decalin this leads to three additional non-bonded interactions of the type found in the *gauche* rotamer of butane compared with the number in *trans*-decalin, with the result that the latter molecule is 13 kJ mol^{-1} more stable than the former. *Trans*-decalin is very rigid, because inversion of one of its cyclohexane rings would convert the equatorial bonds joining its two rings into axial bonds. It is impossible to fuse two cyclohexane rings using axial bonds and for them both to retain chair conformations. At room temperature *cis*-decalin quite readily undergoes a motion which involves the simultaneous inversion of both rings. During this process the equatorial bond joining the two rings is converted to an axial bond and vice-versa.

In adamantane (Fig. 8.24(a)) the lowest frequency vibration occurs at 310 cm^{-1} while bicyclo-[2,2,2]-octane has a vibrational frequency below 20 cm^{-1}. This remarkable difference in rigidity of these two bridged ring systems may be rationalized in terms of the conformations which their constituent cyclohexane rings are forced to adopt. Adamantane is so called because of the similarity of the arrangement of the carbon atoms in this molecule to that found in diamond. The three cyclohexane rings have chair

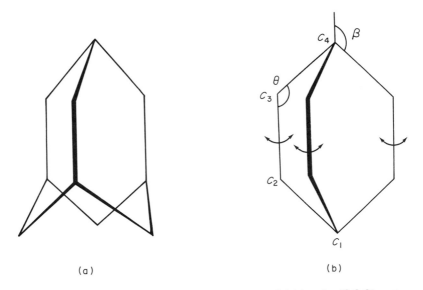

Fig. 8.24   The conformations of (a) adamantane (b) bicyclo-[2,2,2]-octane.

conformations and the molecule is totally free of strain. In bicyclo-[2,2,2]-octane the two cyclohexane rings are forced to adopt full boat conformations. The resulting torsional strain due to eclipsing of hydrogen atoms on the C_2 and C_3 carbon atoms (Fig. 8.24(b)) can be relieved by twisting about the C_2—C_3 bonds. This is counterbalanced by the energy required to distort the angles β and θ and to twist the C_1—C_2 and C_3—C_4 bonds. An electron diffraction study has shown this molecule to have a twisted structure with an angle of 12° between the $C_1C_2C_3$ and $C_2C_3C_4$ planes and a potential barrier of 0·4 kJ mol^{-1} between the two minima. The polar derivative 1-fluoro-bicyclo-[2,2,2]-octane has been studied by microwave spectroscopy with rather similar results.

Further Reading

1. E. L. Eliel, N. L. Allinger, S. T. Angyal and G. A. Morrison. "Conformational Analysis". Interscience, Chichester and New York, 1965.
   Dealing mainly with six-membered rings.
2. J. Laane, Pseudorotation of five-membered rings, Chap. 2 in "Vibrational Spectra and Structure" (J. R. Durig, Ed.), Vol. 1. Dekker, New York, 1972.
   Describes and explains the phenomenon of pseudorotation in five-membered rings.

3. C. S. Blackwell and R. C. Lord. Far-infrared spectra of four-membered ring compounds, Chap. 1 *in* "Vibrational Spectra and Structure" (J. R. Durig, Ed.), Vol. 1. Dekker, New York, 1972.
   Describes the shape and flexibility of four-membered rings.
4. J. Sheridan. Microwave spectroscopy of heterocyclic molecules, Chap. 2, *in* "Physical Methods in Heterocyclic Chemistry" (A. R. Katritzky, Ed.), Vol. 6. Academic Press, London and New York, 1974.
   A comprehensive review of the structure of heterocyclic compounds studied by microwave spectroscopy.
5. H. M. Pickett and H. L. Strauss. *J. Amer. Chem. Soc.* **92**, 7281 (1970).
   Describes the calculation of inversion in six-membered rings.

References

1. G. W. Rathjens, N. R. Freeman, W. D. Gwinn and K. S. Pitzer. *J. Amer. Chem. Soc.* **75**, 5634 (1953).
2. J. K. Kilpatrick, K. S. Pitzer and R. Spitzer. *J. Amer. Chem. Soc.* **69**, 2483 (1947).
3. V. M. Gittins, E. Wyn-Jones and R. F. M. White. "Ring inversion in some six-membered heterocyclic compounds, Chap. 12 *in* "Internal Rotation in Molecules". (W. J. Orville-Thomas, Ed.). Wiley, Chichester and New York, 1974.
4. T. Ueda and T. Shimanouchi. *J. Chem. Phys.* **47**, 4042 (1967).
5. R. P. Bell, *Proc. Roy. Soc. (London)*, **A183**, 328 (1945)
6. D. O. Harris, H. W. Harrington, A. C. Luntz and W. D. Gwinn. *J. Chem. Phys.* **44**, 3467 (1966).
7. F. A. Miller and R. J. Capwell. *Spectrochim Acta.* **26A**, 947 (1971).
8. W. G. Rothschild. *J. Chem. Phys.* **45**, 1214 (1966).
9. H. Kim and W. D. Gwinn. *J. Chem. Phys.* **44**, 865 (1966).
10. C. S. Blackwell, L. A. Carreira, J. R. Durig, J. M. Karriker and R. C. Lord. *J. Chem. Phys.* **56**, 1706 (1972).
11. D. O. Harris, G. G. Engerholm, C. A. Tolman, A. C. Luntz, R. A. Keller, H. Kim and W. D. Gwinn, *J. Chem. Phys.* **50**, 2438 (1969).
12. R. Davidson and P. A. Warsop. *J. Chem. Soc. Faraday Trans. II*, **68**, 1875 (1972).

9
Internal Rotation and Conformational Flexibility in Macromolecules

9.1 Introduction

The physical properties of ethane and simple derivatives of ethane are affected by the occurrence of hindered internal rotation about the central C—C bond, although it would be true to say that the phenomenon plays a relatively minor role in determining the general chemical and physical properties of these molecules. It might be expected therefore, that for very large and complex molecules, the overall molecular properties would be very little affected by internal rotation about single bonds. For certain very important macromolecules, however, the opposite is true. In fact it could be argued that the very existence of life as we know it would not be feasible were it not for the possibility of internal rotation about single bonds.

The structure, shape and function of macromolecules represent aspects of a vast area of study and research, and it would be fruitless to attempt to consider the whole area in detail in a volume such as this, and so the remainder of this chapter will deal with specific examples of three main types of macromolecules, viz: proteins, nucleic acids and carbohydrates, in which internal rotation and molecular flexibility play a very important role. Because of the complexity of these polymers their overall molecular shapes will be determined by the combined effect of a large number of factors including the parameters which govern internal motions such as those described earlier in this book. Clearly, therefore, discussion of the contribution of such forces to the gross molecular conformation in the same detail as given for smaller molecules is not feasible, and consequently the treatment included in this chapter will be very descriptive in nature.

9.2 Proteins

There are several different types of proteins but they all have one feature in common—they are all composed of the same monomeric building blocks, amino acids. Only twenty-three different amino acids have been found which are constituents of proteins and they nearly all have the general formula,

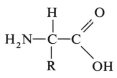

The side group R can range in size from a hydrogen atom (as in glycine) to a bulky derivative, as for example in thyroxine,

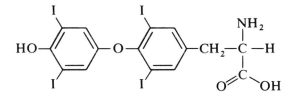

The condensation of two amino acids to form a larger structure may be depicted schematically as follows,

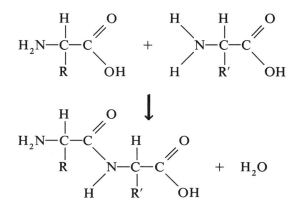

The basic (—NH$_2$) and acidic (—COOH) groups which are present in each amino acid combine, with elimination of water, to form a bond known as the *peptide link*.

In reality the synthesis of most proteins is not such a simple process, but the peptide link is formed nonetheless.

The simplest compounds that possess this important peptide link are amides, and they have attracted a great deal of interest from spectroscopists and other workers interested in molecular structure over the past decade or so. The simplest amide, formamide ($HCONH_2$) has been shown to have a planar molecular framework (see Chapter 7) and NMR spectroscopy and other techniques have shown quite unequivocally that internal rotation about the C—N bond in many simple amides is hindered by a fairly high potential barrier[1] (of the order of 40 kJ mol^{-1} or more) (Fig. 9.1).

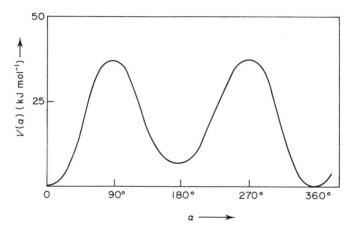

Fig. 9.1   The form of the potential function governing internal rotation about the C—N bond in simple amides (RCONR′R″).

This potential barrier is predominantly twofold in character, with the result that asymmetrically disubstituted amides are able to exist as one of two rotational isomers each with a planar skeletal framework, as shown in Fig. 9.2.

One interesting fact that emerges from studies of monosubstituted amides is that for most substituents the more stable conformation of the molecule is the one in which the free N—H bond lies *trans* to the carbonyl group. This

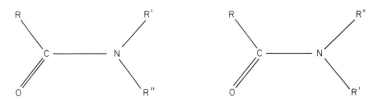

Fig. 9.2   Two possible conformations for a disubstituted amide with the heavy atom skeleton planar or approximately planar.

stable "*trans-amide*" configuration is of course identical to the peptide link found in larger proteins. It is precisely this geometry, with the N—H bond *trans* to the C=O bond (i.e. directed 180° away), that facilitates strong inter-action through hydrogen bonds between neighbouring parts of macro-molecular protein chains.

In certain cyclic amides (lactams) on the other hand the structural require-ments of the ring force the N—H bond to be orientated in the less stable *cis* conformation with respect to the carbonyl bond (Fig. 9.3).

Interestingly, the conformation of the amide group has been shown to be a function of ring size. For lactams of ring size five to eight only the *cis* con-formation is in evidence; for nine-membered rings both *cis* and *trans* (or near *trans*) conformations exist in equilibrium; and for lactams of ring size ten or greater the *trans* amide form again becomes dominant.[2] Thus as the cyclic structures become more flexible (with larger rings) so the more stable *trans* amide conformation is preferred.

As far as the constituent amino acids of a protein are concerned, the key factor which determines the function of the polymeric macromolecule is the

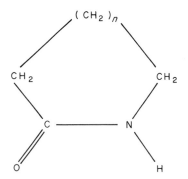

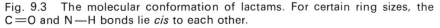

Fig. 9.3   The molecular conformation of lactams. For certain ring sizes, the C=O and N—H bonds lie *cis* to each other.

precise sequence of amino acids in the polymeric chain. Many hundreds of amino acids are joined together in a typical protein. As far as can be established the peptide link retains its planarity (or near planarity) in all macromolecules. Calculations tend to suggest that small deviations (up to about 15°) from planarity would not produce large changes in conformational energy, and there is experimental X-ray evidence to support a deviation from planarity of this order of magnitude in some simple crystalline amino acids.[3] However, the overall molecular shape of the protein polymer depends largely on the nature of the substituent side group R, in particular on the magnitude of the interaction forces, (such as steric hindrance, dipolar, etc.) between the various side groups which force certain sections of the macromolecule to adopt preferred conformations.

Some proteins are stable as helical structures, either as single coils or linked together as twin chains by disulphide bridges. Successive loops of a

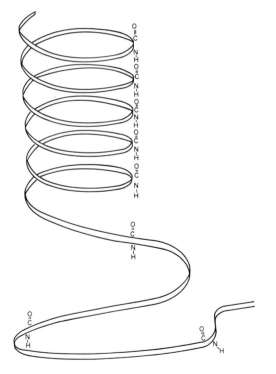

Fig. 9.4   The "helical" and "random coil" structures of poly-γ-benzyl-L-glutamate. The coils of the helix are held together through hydrogen bonds of the type C=O . . . H—N.

protein helix are held firmly together through hydrogen bonds—the *trans* N—H bond of one peptide link interacting with the corresponding C=O bond of the peptide link on the adjacent loop (Fig. 9.4). The importance of the role played by the preferred planar *trans* peptide bond in the intramolecular hydrogen bonding may be illustrated by reference to a synthetic polypeptide (poly-γ-benzyl-L-glutamate), which undergoes a reversible transition between "helical structure" and "random coil structure" under certain conditions. In a hydrogen bonding solvent where strong solute–solvent interactions occur the polymer exists as a random coil, whereas in a non-hydrogen bonding solvent it takes up a helical configuration (Fig. 9.4). In mixed solvents of fixed composition the transition between the two forms takes place over a fairly narrow temperature range. In a strongly hydrogen bonding solvent (e.g. dichloroacetic acid) the peptide is probably firmly associated to the solvent molecules thus restricting the formation of a stable helix structure.[4]

The above helix has about 3·5 monomer units per spiral turn and each is held in place along the spiral by a hydrogen bond. Thus the hydrogen bonds act as a form of "locking" mechanism for the spiral structure—a new loop being formed as soon as the appropriate N—H (or C=O) bond comes within range of its hydrogen bond partner. This "spiral locking" property of the hydrogen bonds (somewhat resembling a zip-fastner effect) is crucially dependent on the respective orientation of the adjoining N—H and C=O bonds. These, in turn, are governed by the geometry of the peptide link. Many protein structures, for example the constituent proteins of hair or wool, adopt a helical structure, one of the most commonly occurring type being the α-helix (Fig. 9.5).

Other well known protein structures, however, are derived from intermolecular hydrogen bonding. An interesting example of such a structure, and one which illustrates the importance of the planar *trans*-CONH peptide conformation in relation to its gross molecular structure and to its physical properties, is silk. Sheet silk is a natural polymer consisting of long polypeptide chains held together laterally by hydrogen bonded N—H ... O groups. Within each silk polymer strand the *trans* amide arrangement, which arises as a result of the potential forces hindering internal rotation about the C—N bond, ensures that each peptide group is linked to other peptide groups on both sides via N—H ... O bonds. The result of this is that the final structure represents a continuous sheet of zig–zag polymer chains (Fig. 9.6).

In order to facilitate both intermolecular hydrogen bonding and the tetrahedral symmetry of each carbon atom the extended polypeptide chain has a "pleated" β-sheet structure (Fig. 9.7).

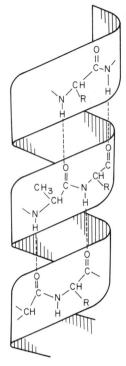

Fig. 9.5   The helical structure of the α-helix.

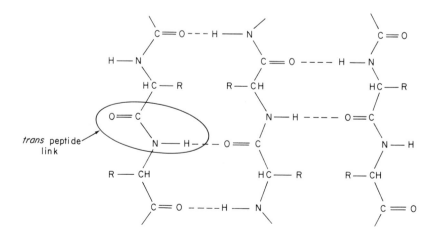

Fig. 9.6   The structure of sheet silk showing how the orientation of the C=O and N—H groups are crucial to the overall structure.

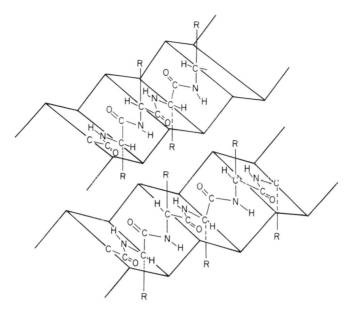

Fig. 9.7   "Pleat" structure of sheet silk.

The smaller the side-group **R**, the more efficient the stacking of these pleated silk sheets on each other, and in natural silk many of the side groups are hydrogen atoms, thus helping to give silk its characteristic physical properties.

9.3 Nucleic Acids

Nucleic acids are found in the chromosomes of living cells. These chromosomes play an important role in controlling the genetic information inherent in a living organism, and it is believed that the key constituents in chromosomes with respect to this genetic coding are the nucleic acids. Nucleic acids are comprised of three basic components; a backbone of sugar molecules linked by means of bridging phosphate groups, with an organic base (one of four different nitrogen-containing heterocyclic compounds) attached to each sugar molecule. Deoxyribonucleic acid (DNA), which is the major component of chromosomes has a basic skeletal structure as shown in Fig. 9.8.

Extensive research work has shown that a single unit of DNA normally comprises two polymeric chains coiled together in fairly rigid helical shape.

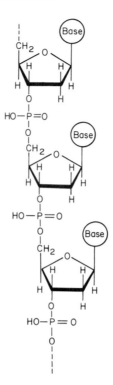

Fig. 9.8   The skeletal structure of deoxyribonucleic acid (DNA). One of four organic bases may be utilized; adenine, thymine, guanine and cytosine.

The two chains are linked together through the nitrogen containing bases. A base on one helix is joined by hydrogen bonds to a base at an appropriate position on the other helix, and all the bases are thought to be connected in this way throughout the double helix (Fig. 9.9). This specific type of pairing requires the bases to lie in the same plane, approximately perpendicular to the axis of the double helix. The entire DNA structure is feasible only by virtue of the internal degree of freedom which atomic groups possess enabling them to rotate, albeit against a hindering potential barrier, about single bonds. In this instance the orientation of the bases is governed largely by the form of the potential energy hindering internal rotation about the C (sugar)—N(base) bond. This hindering potential ensures that the hydrogen bridging bases are able to align together in the correct manner. This internal flexibility is also likely to be of paramount importance in the duplication of DNA chains, whereby one helix uncoils from its partner and the two then act as templates to generate two other helical chains.

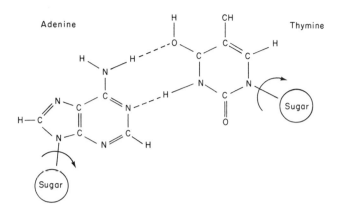

Fig. 9.9   The role of hydrogen bonding in linking up the two helices in DNA through a purine and pyridine type base on corresponding turns of the separate helices.

It is interesting to note that the C=O and N—H groups in thymine which participate in specific intermolecular hydrogen bridges with corresponding groups on adeninine are arranged in a *cis* configuration. In this instance the rigidity of the ring structure in thymine forces these groups to be locked in this orientation.

9.4 Carbohydrates

Carbohydrate polymers are natural polymers that occur widely in foodstuffs and in plant and animal life. The predominant monomeric units are known as monosaccharides or sugars, and unless the carbohydrate polymer contains substantial proportions of non-carbohydrate material (e.g. nitrogen and phosphorous containing compounds) they are known as polysaccharides. Whereas the structure and properties of a protein macromolecule depend largely on the amino acid sequence, for polysaccharides the same mono-meric sugar units can link up in a variety of different ways thus producing for example such diverse materials as cellulose or starch from monomeric glucose. Thus although they represent less complex structures than the proteins, in the sense that the variety of different monomers contributing to a polysaccharide structure is generally less than for a protein, in principle there is a wider possible variety of polysaccharides than polypeptides. In reality, polysaccharides lack the molecular diversity of proteins.

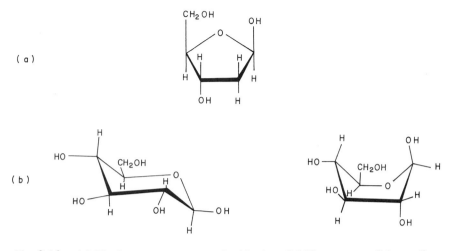

Fig. 9.10   (a) The furanose monosaccharide ring. (b) The two possible configurations of the pyranose ring structure.

Monosaccharide molecules are usually found to exist as five- or six-membered rings (furanose or pyranose rings respectively). The five-membered rings are approximately planar (Fig. 9.10(a)) but the six-membered pyranose structures are puckered and have very similar conformational properties to cyclohexane and derivatives, i.e. they may exist in a chair or boat form (Fig. 9.10(b)). Just as for cyclohexane derivatives, pyranose sugars are more stable in the chair conformation, and as far as is possible the bulky side groups tend to occupy equatorial positions. It may be interesting to note at this juncture that the sugar units involved in the structure of DNA are all pentoses and form five-membered, nearly planar, furanose rings.

Detailed structural studies of polysaccharides are few owing to the non-crystalline nature of carbohydrates, but some interesting deductions have been made concerning certain selected compounds. ι-Carrageenan[5] is a polymer of the type $(-A-B-)_n$ where the monomeric units A and B are derived, respectively, from β-D-galactose-4-sulphate and 3,6 anhydro-α-D-galactose (Fig. 9.11(a)). It is an important constituent of the protective jelly layers of certain coastal seaweeds. Indeed, the unique property of carrageenan to form stiff gels, even in solutions as dilute as 1 part carrageenan to 99 parts of water, has intrigued many people. Such gels are now believed to form because of the presence of a skeletal framework of carrageenan polymers partially arranged as double helices between different polymer threads (Fig. 9.11(b)).

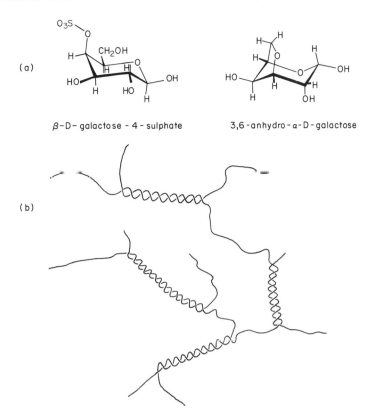

Fig. 9.11   (a) The configurations of β-D-galactose-4-sulphate and 3,6-anhydro-α-D-galactose. (b) Showing the double helices which can exist between the various carbohydrate polymer strands.

The double helices are held together by O—H . . . O hydrogen bonds which are aligned perpendicular to the axis of each helix. It is thought highly probable that such helical structures are formed in solution by a pair of polymer chains randomly forming the first turn. Once the first turn is formed the hydrogen bonds hold the structure until other turns are completed. The orientation of these strategically placed hydrogen bonds is a direct function of the nature and shape of the potential hindering internal rotation about the C—O bonds which link the monomeric galactose units together (Fig. 9.12). The inhomogeneity or "kinks" in the double helices are believed to be caused by the occasional presence of a third type of sugar monomer.

Gels of the type that carrageenan forms have two characteristic temperatures, melting and setting temperatures, which can differ considerably. The

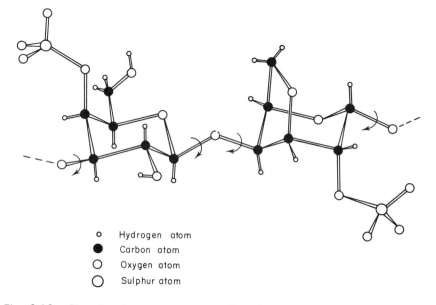

o	Hydrogen atom
●	Carbon atom
○	Oxygen atom
○	Sulphur atom

Fig. 9.12   Showing how two monomeric galactose units link together in a carbohydrate polymer.

melting point (which is often fairly low) characterizes that point when the helices are not stable enough to retain the polymeric skeletal backbone. This is almost certainly due to the fact that oscillatory motions within the chains prevent the retention of the hydrogen bonded links. Upon recooling, the gel will not reform until a temperature below the melting point is reached (setting point). Thermal energy ensures that the oscillatory motions of the various groups about the single bonds prevent the reforming of the initial hydrogen bonds, and thus the temperature has to be below the setting point before the resetting mechanism becomes operational.

Some interesting results pertinent to the conformational features of poly-saccharides have been derived from computations carried out on the conformation of disaccharides such as cellobiose[6] (Fig. 9.13). By assuming fixed ring conformations for the two monosaccharide units with standard values for the bond lengths and bond angles (including the central C—O—C angle), the internuclear distances between the various atoms located on the different sugar units, and the van der Waals energy between these atoms were computed for various values of the torsional angles ϕ and ψ (Fig. 9.13). It was found that in order to have a sterically non-hindered disaccharide structure only a small range of values of these angles was permissible.

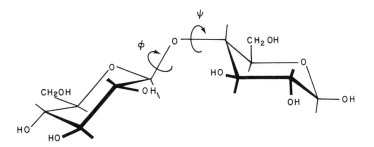

Fig. 9.13   The structure and conformation of cellobiose.

Significantly, the experimental X-ray data for cellobiose falls within this restricted range.

9.5 Conclusion

In this brief account of the role of hindered internal rotation in polymeric compounds, examples have been considered from ordered polymers. Much interesting work has also been carried out on the shapes of random polymers. These may be computed using statistical arguments and employing input parameters (e.g. bond lengths and angles, internal potential barriers) derived for the monomeric units. Indeed, with increasing computer facilities, the time may be close that quite accurate structural data may be derived for macromolecules based on the experimental findings for the constituent monomeric entitities.

In conclusion it should be noted that the considerable interest that scientists have in the details of the molecular structure of simple compounds such as amides may be appreciated when one considers that the same forces which dictate the preferred conformation of N-methyl acetamide probably contribute substantially to the structure of the α-helix found in proteins.

Further Reading

The current interest in and the importance attached to the conformational behaviour of macromolecules may be gauged by the fact that practically all new books on biochemistry include this topic as a major section.

1. The interested reader is directed towards a special volume dedicated to Linus Pauling:

A. Rich and N. Davidson (Eds). "Structural Chemistry and Molecular Biology". W. H. Freeman, San Francisco 1968.
Several chapters within this volume discuss aspects of the conformations of macro-molecules.
2. D. A. Rees. "The Shapes of Molecules". Oliver and Boyd, Edinburgh, 1967.
An interesting monograph dealing with the size and shapes of carbohydrates.
3. L. Pauling and R. Hayward. "The Architecture of Molecules". W. H. Freeman, San Francisco, 1964.
Beautifully illustrated pictures of molecular models, some showing conformational properties.

References

1. H. Kessler. *Angew. Chem.* (Internat. Edit.), **9**, 219 (1970).
2. (a) R. Huisgen and H. Walz. *Chem. Ber.* **89**, 2616 (1956).
   (b) R. Huisgen, H. Brade, H. Walz and I. Glogger. *Chem. Ber.***90**, 1437 (1957).
3. M. Perricaudet and A. Pullman. *Int. J. Peptide Protein Res.* **5**, 99 (1973).
4. M. Davies. *J. Chem. Ed.* **46**, 17 (1969).
5. D. A. Rees. *Science*, **6**, 47 (1970).
6. D. A. Rees and R. J. Skerrett. *Carbohydrate Res.* **7**, 334 (1968).

Index